WILLIAM SMITH

The Last Frontier

How We'll Save or Lose the Solar System

First edition

This book was professionally typeset on Reedsy.
Find out more at reedsy.com

Contents

Preface

Why I Wrote This Book

I will never forget the night I realized we were making the same mistake twice.

It was a clear evening in late 2019, and I was standing on a cold hillside outside the city, wrapped in a jacket that was not quite warm enough, staring up at a sky I had loved since childhood. I had driven out there to escape light pollution and spend a few quiet hours with a small telescope I had owned for years. The ritual was familiar and comforting. Align the mount. Let my eyes adjust. Breathe. Look up.

Then the sky began to move in a way I had never seen before.

A bright line cut across my field of view. Then another. And another. Within minutes, faint streaks were appearing everywhere, silent and relentless, sliding through constellations I knew by heart. At first I thought something was wrong with my equipment. Then I realized nothing was wrong at all. The sky had changed.

I remember lowering the telescope and just standing there, feeling a strange mix of awe and unease. These were satellites, part of a new generation of mega constellations designed to blanket the planet with connectivity. From an engineering perspective, it was impressive. From a human perspective, it was unsettling. From an environmental perspective, it felt painfully familiar.

That was the moment it clicked.

We were treating space exactly the way we treated Earth. Extract first. Deploy fast. Optimize for profit and speed. Worry about consequences later. It was the same pattern that had given us climate change, collapsing fisheries, poisoned rivers, plastic filled oceans, and mass extinction. The only difference was the setting. This time, the frontier was orbital space, the Moon, Mars, and worlds we had barely begun to understand.

What shocked me most was not what I saw in the sky, but how little anyone seemed to be talking about it.

The personal journey

I am not an astronaut, and I am not a rocket scientist. I am a researcher and writer who has spent years trying to understand how large systems fail and how, sometimes, they are saved. My background is in environmental research and science communication, work that has kept me close to climate science, conservation efforts, and the long, frustrating history of humanity realizing a problem far too late.

I grew up loving space. I devoured books about planets and stars, built model rockets, and stayed up late to watch grainy footage of launches and landings. Space represented curiosity without borders, cooperation across nations, and a sense that humanity could be better than its worst instincts. For a long time, space felt clean in a way Earth no longer did. It felt untouched, almost innocent.

As my career pulled me deeper into environmental work, that innocence began to crack. I learned how quickly ecosystems can unravel once certain thresholds are crossed. I learned how often early warnings are ignored because the damage is not yet visible or politically inconvenient. I learned how the phrase we will fix it later almost always translates into we will not fix it at all.

When I started paying closer attention to space, I saw the same warning signs everywhere. Launch rates accelerating year after year. Orbital debris counts rising steadily. Planetary protection protocols quietly being reconsidered in the name of efficiency. Private companies talking openly about mining asteroids and settling Mars with far more confidence than caution.

The turning point came when I attended a space policy conference and listened to a discussion about orbital debris. The room was full of brilliant people. Engineers, policymakers, industry representatives. The risks were laid out clearly. Collision cascades. Loss of critical satellite infrastructure. Decades or centuries of orbital environments rendered unusable. Everyone understood the science. Everyone agreed the danger was real.

Then the conversation moved on.

There was no urgency in the room, no sense that this was a crisis demanding immediate, coordinated action. It was treated as a technical challenge to be managed someday, preferably by someone else. I left that conference with a sinking feeling I recognized all too well. I had seen this before, in climate meetings decades earlier, when the data was already screaming and the response was still polite concern.

I went looking for a book that pulled all of this together. A book that explained not just space debris or planetary contamination in isolation, but the pattern connecting them to Earth's environmental history. A book that treated space not as an abstract playground for technology, but as an environment subject to the same limits, vulnerabilities, and ethical questions as any other. A book that did not just describe the problem, but explored solutions and highlighted the people fighting to make them real.

That book did not exist.

So I wrote it.

Why this book matters now

This is not a book about distant futures or speculative centuries. It is about decisions being made right now, in the 2020s and early 2030s, decisions that will lock in trajectories for generations. Every year, launch numbers increase. Satellite constellations grow denser. Plans for crewed Mars missions move from science fiction to concrete timelines. Resource extraction proposals shift from theory to investment decks.

We are at a fork in the road.

History has shown us how this plays out if we are not careful. With climate change, scientists were raising alarms as early as the 1980s. The physics was clear. The consequences were predictable. Yet serious global action was delayed for decades. By the time the crisis felt undeniable to the public, we were already in damage control mode. The cost of inaction compounded, and the window for easy solutions closed.

Space offers us something we have never had before. A chance to do it right from the beginning. To put protections in place before ecosystems are damaged, not after. To design governance systems before exploitation becomes entrenched. To choose restraint where restraint matters most.

That chance is narrow, and it is closing.

What happens in orbit over the next ten to fifteen years will determine whether low Earth orbit remains usable or becomes a minefield. What we do on Mars in the first few human missions will determine whether we preserve its scientific record or contaminate it beyond recovery. How we approach the Moon and asteroids will determine whether they become shared scientific resources or industrial sacrifice zones.

This book had to be written now because awareness still lags behind reality. Because many people who care deeply about the environment have no idea that space is already facing environmental pressures. Because many people who love space exploration have been told, incorrectly, that sustainability and regulation are threats rather than safeguards. Because the conversation is fragmented, technical, and too often hidden from public view.

Most importantly, it had to be written now because choices made in ignorance are still choices, and ignorance is no longer an excuse.

What makes this book different

There are excellent books about space exploration, and there are powerful books about environmentalism. There are academic papers that dive deeply into orbital mechanics or planetary protection protocols. This book sits in the space between them, because that is where the problem lives.

This book connects Earth and space, showing how the same dynamics of extraction, inequality, and delayed accountability repeat across environments. It is written for people without advanced degrees, because the future of space should not be decided only by specialists. It does not stop at diagnosis, because understanding the problem without seeing a path forward breeds despair. It centers people, because movements are built by humans, not abstractions.

Most of all, it is written to make you care.

If you finish this book knowing more facts but feeling indifferent, I have failed. If you finish feeling overwhelmed and powerless, I have failed. If you finish thinking that space environmentalism is someone else's problem, I have failed.

If you finish believing that we can do better, that we must do better, and that you have a role to play, then this book has done what it was meant to do.

Who this book is for

I wrote this book for people who love space and do not want to see it ruined by the same short term thinking that damaged Earth. Loving space does not mean defending every launch or dismissing every regulation. It means caring enough to protect it.

I wrote this book for environmentalists who have spent years fighting for forests, oceans, and the climate, and who may not yet realize that the environment now includes orbital space and other worlds. Environmental protection does not stop at the edge of the atmosphere.

I wrote this book for students and young people, because you will live with the consequences of decisions being made now. You deserve to understand what is at stake and to have the tools to demand better.

I wrote this book for curious readers who sense that something is changing in space and want a clear, honest explanation of what that change means.

I wrote this book for skeptics who believe space is infinite and therefore indestructible, or who think we should focus on Earth first. I invite you to test those assumptions against evidence and history.

Ultimately, I wrote this book for anyone who believes that the future should be shaped deliberately rather than stumbled into. You do not need to be a scientist or an activist. You only need to care about what kind of future we are building.

How to use this book

This book is structured as a journey. It begins with what Earth's environmental history can teach us, then moves outward to the Moon, Mars, asteroids, and ocean worlds. It examines the threats, explores the ethical questions, and lays out both technological and governance based solutions. It then asks you to imagine two possible futures and introduces the movement working to make the better one real.

You can read it straight through, which is how it was designed. Each section builds on the last. If you are short on time, you can focus on the Prologue, the opening chapters, the climax, and the Epilogue to grasp the core argument. If you are looking for action, the later chapters provide concrete ways to get involved.

Throughout the book, key ideas are emphasized, sources are provided for deeper exploration, and questions are posed intentionally. They are there to slow you down, not to quiz you.

A final personal note

Writing this book changed me. I began it worried and frustrated, convinced the problems were real and unsure whether solutions could gain traction. I finished it more hopeful than I expected. Not because the challenges are small, but because the people working on them are real, dedicated, and growing in number.

The risks to the space environment are serious and urgent. But they are not inevitable. The solutions exist. The movement exists. What happens next depends on whether awareness turns into action.

Any mistakes in this book are mine. Any oversimplifications were made in the service of clarity, not dismissal of complexity. The space environment is intricate, and no single volume can capture all of it. My goal has been honesty, accessibility, and urgency without panic.

Fifty years from now, people will look back at this moment and ask what we did when we realized space was vulnerable. They will ask whether we repeated our mistakes or learned from them.

Which question they ask depends on us.

On you.

Turn the page. Let me show you what is at stake, and what we can still do about it.

Prologue

Dawn on Mars arrived without drama. The sun lifted itself above the horizon as a muted disc, its light thinned by distance and dust, spreading slowly across the plains and filtering into the reinforced windows of the Mars Historical Archive. Inside, Dr. Elena Vasquez stood motionless, watching the light creep across the floor in pale bands that caught on the edges of consoles and holographic projectors. Outside the glass, red dust clung stubbornly to the window seams, worked into every groove by decades of storms, a reminder that Mars never offered beauty without resistance.

The archive breathed around her. The sound was subtle but constant, a low mechanical hum that came from the artificial gravity systems compensating for the planet's weakness. At thirty eight percent of Earth's gravity, objects seemed to hesitate before committing to motion, and Elena's body, shaped by this world since birth, accepted the light pull as natural. Earth gravity, which she had experienced only during childhood visits, had pressed down on her then with an unfamiliar insistence, making her bones ache and her lungs feel crowded, as if the planet itself demanded acknowledgment of its mass.

She inhaled slowly. The air was thin but clean, carefully regulated and endlessly recycled. It carried the faint scent of metal, filtration mesh, and the charged sharpness of holographic emitters warming into readiness. Beneath her boots, the floor panels were cool, their composite surface hiding layers of conduits that carried power and preserved memory. Everything in the archive existed to endure.

Elena preferred this hour before the archive filled with voices. Before young colonists arrived with restless energy and half formed certainty. Before history was asked to explain itself. In the early morning quiet, the records felt less like testimony and more like companions waiting patiently to be heard.

She rested her hand on the main console, her fingers careful and deliberate, shaped by decades of precision and restraint. Light unfolded in response, rising into the air as layered holograms that shimmered softly. Dates, images, sound fragments, and data streams arranged themselves into a familiar architecture. Humanity, translated into patterns of light.

In a few hours, students would stand where she stood now. They were children of Mars, born under domes, raised with the knowledge that air was engineered, water was counted, and nothing was truly infinite. Earth, to them, was a lesson rather than a place. Elena had been asked to tell them why their world functioned the way it did, and why the rules that governed their lives were written with such caution.

The answer waited inside the archive.

The file glowed at the center of the display, restrained but insistent. **2025 to 2075: Divergence Records**. Every archivist recognized the title. Many avoided opening it for longer than necessary. Elena had spent her life returning to it, drawn by the knowledge that everything that followed depended on what lay inside.

She expanded the timeline with a slow movement of her hand. The holograms responded instantly, branching outward in luminous threads. One path divided into two, then subdivided again, each branch representing projections, probabilities, and consequences. The structure resembled a river delta seen from orbit, or the neural maps Elena had once studied while trying to understand how memory itself chose what to preserve. Choice, rendered visible.

"This is where certainty ended," she said quietly, though no one else was present to hear it.

The year 2025 pulsed faintly at the center. News fragments appeared around it, suspended in light. Climate conferences promised cooperation. Private corporations announced ambitious launch schedules. Governments debated regulation while approving exploratory contracts. Space was framed as opportunity, as solution, as escape. The language was confident and polished, and it carried an undertone Elena now recognized as denial.

She remembered that era through the lens of childhood. She had been living on Mars already, in the early habitats where every kilogram shipped from Earth was logged and justified. When her parents took her back to Earth during rotation breaks, the abundance had stunned her. Skies stretched endlessly overhead, oceans extended beyond the horizon, and cities glowed wastefully at night. Even then, she had noticed the contradiction between urgent conversations about planetary limits and the casual excess that followed them.

As she advanced the timeline into the 2030s, the divergence sharpened. Two distinct philosophies of expansion took shape, each supported by its own data models and economic forecasts. The archive displayed them side by side, their early differences subtle enough to be dismissed by those who preferred simplicity. Resource extraction proposals expanded. Environmental assessments were commissioned, revised, and delayed. Early drafts of what would later become the Interplanetary Environmental Accords appeared briefly, buried beneath lobbying correspondence and procedural deferrals.

Elena leaned closer to the display, her body drifting slightly with the reduced gravity before settling again. She could feel the archive's hum through the soles of her boots, steady and patient. This was the moment when humanity possessed both the knowledge and the capacity to choose differently, and still argued over whether responsibility extended beyond Earth.

The records did not editorialize. They simply unfolded.

The archive labeled the first projection without ceremony. Timeline A.

The images that followed were stark. Worlds stripped faster than models had predicted. Lunar regolith scarred by unregulated extraction. Asteroid mining operations that prioritized yield over containment, leaving clouds of debris that rendered entire orbital corridors hazardous. Early Martian colonies expanded too quickly, pushing life support systems beyond redundancy and resilience. When failures occurred, they cascaded.

Elena watched as simulated ecosystems collapsed under cumulative strain. Soil experiments failed as imported biomes degraded. Water reserves were contaminated by shortcuts taken in waste processing. Colonies that had been advertised as self sustaining required constant resupply, draining Earth even further. The promise of space as relief became a multiplier of damage.

Earth itself did not disappear in these projections, but it diminished. Coastal cities withdrew behind barriers that eventually failed. Agricultural zones shrank under unpredictable climates. Political alliances fractured as resources tightened. Space had not saved humanity. It had merely extended the pattern of consumption into new territory.

She felt the familiar tightening in her chest as the archive displayed the final markers of this path. Abandoned stations. Silent domes. Data logs ending abruptly. Human presence scattered and fragile, stretched too thin across too many damaged worlds.

The archive did not linger. It moved on, as if aware that this vision was not meant to be endured indefinitely.

Timeline B emerged more slowly, its early years almost indistinguishable from the first. The same launches. The same debates. The same tension between ambition and restraint. What changed were the margins.

Environmental thresholds were enforced rather than negotiated away. Extraction quotas were tied to restoration requirements. Closed loop systems were treated as ethical baselines rather than experimental luxuries. Expansion slowed, and in slowing, it stabilized.

The archive showed habitats designed around limitation rather than dominance. Martian settlements grew inward before outward, prioritizing resilience over scale. Lunar operations focused on infrastructure rather than profit. Asteroid resources were harvested with containment fields that cost more but preserved orbital safety. The Interplanetary Environmental Accords appeared again, this time ratified, revised, and enforced.

Elena watched cities form that resembled ecosystems more than machines. Waste streams became inputs. Energy systems adapted to local conditions rather than imposing imported solutions. Earth's decline did not reverse, but it stabilized, given time to recover as pressure eased.

Human presence across the solar system in this timeline was quieter. Fewer headlines celebrated speed or scale. Instead, the data reflected continuity. Colonies endured. Knowledge accumulated. Children were born into systems designed to last longer than any single generation.

Hope, Elena knew, was not spectacle. It was maintenance.

She closed both projections with a gentle motion, allowing the branching light to fold back into the central timeline. The archive returned to stillness, its hum unchanged.

Which future had become real was not written on the surface of the records.

The archive preserved possibilities alongside outcomes, refusing to simplify the past into inevitability. Elena had learned long ago that history was not a single line but a landscape shaped by countless decisions, many of them made quietly, without ceremony.

She turned away from the console and looked out through the reinforced windows again. The sun had climbed higher now, casting longer shadows across the dust outside. Mars remained indifferent, as it always had. It offered no approval and no forgiveness. It simply existed, waiting to see how carefully it would be treated.

The students would arrive soon. They would expect answers, or at least clarity. Elena knew better than to offer certainty too easily. The power of this moment lay not in revealing outcomes, but in understanding how fragile the turning point had been.

History, she believed, was not meant to absolve or accuse. It was meant to orient.

She gathered the presentation markers, aligning the holographic layers she would show first. Her hands moved with practiced calm, though memory stirred beneath the surface. She thought of Earth as she had last seen it, blue and cloud wrapped, already bearing scars that had once seemed abstract. She thought of the long transfer ships that followed, carrying not just people and cargo, but the weight of unfinished decisions.

When the doors opened and young voices filled the archive, she would begin at the beginning. Not with triumph. Not with disaster. With a choice.

She straightened, feeling the light pull of Mars beneath her feet, and allowed herself one last glance at the sleeping records behind her.

"Let me show you how we got here," she said softly, preparing to speak the truth at last. "Starting with the day we broke Earth."

Introduction

The Cosmic Turning Point

On July 20, 1969, Neil Armstrong stepped onto the Moon and spoke words that have echoed for more than half a century. He called it one small step for a human being and one giant leap for humankind. In a literal sense, he was describing distance measured in meters. In every other sense, he was describing a transformation of our species. For the first time in our history, we had left the planet that shaped us. We had reached out across space and touched another world.

What Armstrong could not have known, and what few people fully appreciated at the time, was that this leap carried with it a choice that would unfold slowly, quietly, and largely outside public attention. Becoming a space faring species was not only a technical achievement. It was a moral and environmental turning point. It forced a question that humanity has never had to answer at a planetary scale before. What kind of space faring species will we be.

Will we carry into the cosmos our best qualities, including curiosity guided by humility, cooperation across borders, patience, and stewardship rooted in long term thinking. Or will we export our worst habits, including extraction without restraint, short sighted profit seeking, disregard for consequences, and the assumption that damage can always be fixed later.

More than five decades after that first footprint was pressed into lunar dust, we are still answering that question. The answer is not written in speeches or mission statements. It is written in launch schedules, regulatory gaps, investment flows, and quiet technical decisions that rarely make headlines. The answer will determine not only the future of space exploration, but the future of the systems on Earth that increasingly depend on space remaining usable and stable.

This book begins from a simple but unsettling premise. Humanity is approaching a cosmic turning point. The decisions made in the coming decades will either set us on a path toward sustainable, responsible presence beyond Earth or lock us into patterns of damage that mirror the environmental crises we already face at home. The stakes are higher than they appear, and the window for shaping the outcome is narrower than most people realize.

The pattern we keep repeating

Human history is filled with moments of discovery that begin in wonder and end in regret. The pattern is so familiar that it often fades into the background, accepted as inevitable rather than examined as a choice. When humans encounter a new place, whether a forest, an island, a continent, or a previously unreachable ecosystem, the sequence rarely changes.

It begins with discovery. A place is encountered and described as new, even when it has been inhabited or shaped for millennia. The dominant response is awe. Stories are told. Maps are drawn. The unknown feels full of promise.

Discovery is followed by exploration. People study the place, often carefully and respectfully at first. Naturalists catalogue species. Geographers chart coastlines and rivers. Scientists ask questions. At this stage, the damage is usually limited, and the prevailing narrative is one of learning and progress.

Then comes exploitation. Attention shifts from understanding to opportunity. Resources are identified. Timber, minerals, fertile land, strategic positions, and sources of energy become the focus. The language changes subtly but decisively. The place is no longer valued for what it is, but for what it can provide.

Extraction follows quickly. Technologies are deployed to remove value at increasing speed and scale. The pace of removal accelerates as competition intensifies. What was once abundant is treated as limitless, even when early warnings suggest otherwise.

Environmental destruction emerges as a consequence rather than a deliberate goal. Forests disappear. Species vanish. Waters are polluted. Soils are depleted. Because the damage accumulates gradually, it is often invisible until it crosses a threshold that cannot be undone.

Inequality is woven through every stage of this process. Those who benefit most from extraction are rarely those who bear the costs. Profits flow upward and outward, while local communities and future generations inherit degraded environments and diminished options.

Eventually, regret sets in. Laws are passed. Protected areas are created. Restoration efforts are launched. These steps are often sincere and sometimes effective, but they arrive after the most valuable aspects of the environment have already been lost. The refrain is familiar. We did not know. We did not act in time.

This pattern has repeated across eras and regions. Medieval Europe stripped its forests for fuel and ships, reshaping landscapes and economies. Colonization of the Americas brought ecological transformation and human catastrophe on an unprecedented scale. The industrial revolution delivered wealth and power while locking the world into fossil fuel dependence. The modern era has given us climate change, mass extinction, and planetary scale pollution. Each time, the story is framed as progress, and each time the costs are discounted until they become unavoidable.

The persistence of this pattern is not mysterious. Short term incentives consistently outweigh long term consequences. Political systems reward decisions that pay off within election cycles rather than ecological timescales. Economic structures allow costs to be externalized onto people without power and onto generations not yet born. Those who profit from extraction are often insulated from the damage it causes. Awareness lags action, and by the time the full extent of harm is visible, infrastructure and vested interests make change difficult.

Perhaps most importantly, humans repeatedly convince themselves that this time will be different. We believe that better technology, greater knowledge, or improved governance will allow us to avoid past mistakes without fundamentally changing our behavior. History suggests otherwise.

Now, humanity is extending this pattern beyond Earth. The early stages are already visible. We have discovered and explored the Moon, Mars, asteroids, and the moons of the outer planets. Scientific missions have revealed environments of astonishing complexity and value. At the same time, commercial activity in space is accelerating rapidly. Private companies are launching thousands of satellites, planning resource extraction, and proposing permanent settlements on other worlds.

We are moving from exploration to exploitation, and in some cases to extraction, faster than we are building the ethical and regulatory frameworks needed to guide those activities. Orbital debris is accumulating. Planetary contamination risks are increasing. Inequality is evident in who has access to space and who stands to benefit from its resources.

The critical question is whether we can interrupt the pattern this time. For the first time in history, we have the advantage of foresight. We can see the consequences of unrestrained exploitation before they fully unfold. We know what happens when regulation follows damage rather than precedes it. We know how difficult it is to reverse environmental harm once thresholds are crossed.

The opportunity before us is not to avoid change, but to choose a different sequence. Protection before extraction. Governance before exploitation. Humility before hubris. Whether we seize that opportunity remains uncertain.

Why the space environment matters

To many people, concern about the space environment initially sounds misplaced. Space feels vast, empty, and resilient. Compared to the fragility of forests or coral reefs, the vacuum above Earth seems immune to damage. This intuition is understandable, but it is wrong in ways that matter deeply.

Space is not infinite for human use. The regions of space that matter most to us are narrow and specific. Low Earth orbit is a finite shell extending only a few thousand kilometers above the planet. It is where most satellites operate, including those that enable navigation, communication, weather forecasting, and Earth observation. Once crowded with debris, this region can become hazardous or unusable for decades or centuries.

Similarly, the Moon has a limited number of locations of exceptional scientific and strategic value. Permanently shadowed craters at the lunar poles contain ice that records billions of years of solar system history. Mars preserves a geological archive that may hold evidence of ancient life. The moons of the outer planets may host oceans that could answer one of humanity's most profound questions about whether life exists elsewhere. These environments are not interchangeable, and damage to them cannot be undone.

Another common objection is that humanity should focus on fixing Earth before worrying about space. This framing presents a false choice. Space and Earth are already deeply interconnected. Modern civilization depends on satellites for navigation, communication, disaster response, and climate monitoring. Damage to space infrastructure directly undermines our ability to protect and manage Earth's environment.

Beyond practicality, the way we treat space reflects our capacity to learn from experience. If we cannot apply lessons from Earth's environmental crises to a new domain where the warning signs are visible and the stakes are clear, it raises troubling questions about our ability to govern ourselves responsibly anywhere.

It is also tempting to view space environmental issues as purely technical problems. From this perspective, debris can be removed with better technology, contamination can be managed with improved sterilization, and resource extraction can be optimized with engineering solutions. Technology is necessary, but it is not sufficient. The hardest questions are not about how to do things, but about who decides, who pays, who benefits, and who bears the risks. These are political, economic, and ethical questions, not just technical ones.

The stakes extend beyond infrastructure and science. They touch on ethics and identity. Becoming a space faring species is not simply about expanding territory. It is about deciding what responsibilities come with that expansion. Do we see other worlds as raw material to be consumed, or as environments with intrinsic value. Do we consider the rights of future generations when making irreversible changes. Do we recognize that some questions, such as whether life exists elsewhere, can only be answered once, and that careless action can destroy the evidence forever.

In this sense, the space environment is a test of human maturity. It asks whether we can act with restraint when restraint is in our long term interest.

It asks whether cooperation can prevail over competition in a domain where unilateral action can harm everyone. It asks whether we can value knowledge, beauty, and possibility alongside profit and power.

The central thesis of this book

This book advances a set of arguments that build toward a single conclusion. Humanity stands at a crossroads in its relationship with space, and the path we choose will shape centuries to come.

First, the book argues that we are in the early stages of exporting Earth's environmental crisis to space. The same dynamics that drove deforestation, pollution, and climate change are visible in orbital debris accumulation, planetary contamination risks, and unregulated resource extraction. The actors may differ, but the incentives and failures are strikingly similar.

Second, the book argues that damage to the space environment has direct and indirect consequences for Earth. A debris cascade in low Earth orbit would disrupt systems that underpin global navigation, communication, and disaster response. The loss of scientific knowledge about other worlds would be a loss for all humanity. The precedents we set in space will influence how future environments are treated, both off world and on Earth.

Third, the book argues that the window for meaningful action is limited. The period from the mid 2020s through the mid 2040s is likely to determine whether space becomes a sustainable domain or a degraded one. Once certain thresholds are crossed, such as an uncontrollable debris cascade or widespread planetary contamination, the consequences become effectively permanent on human timescales.

Fourth, the book argues that solutions exist. Technologies for debris mitigation, contamination prevention, and sustainable resource use are real and improving. Governance models drawn from international cooperation on Earth provide useful precedents. Ethical frameworks rooted in long term thinking and indigenous wisdom offer alternatives to short term exploitation. What is lacking is not knowledge, but collective will.

Finally, the book argues that two broad futures are possible. One leads toward environmental degradation, conflict, and foreclosed possibilities. The other leads toward a sustainable human presence that balances exploration with protection. Which future emerges is not predetermined. It will be shaped by choices made now, by institutions and individuals alike.

Space environmentalism, as presented here, is not opposed to exploration, technology, or progress. It is an argument for doing these things wisely. It insists that exploration without stewardship is not progress, and that technological power without ethical restraint is a liability rather than an achievement.

Key terms and concepts

To navigate the arguments that follow, it is helpful to clarify several terms that will appear throughout the book.

The space environment refers to all regions beyond Earth's atmosphere that are affected by human activity, including orbital space, the Moon, planets, asteroids, and the moons of other planets. These are not empty voids, but environments with physical, scientific, and sometimes biological characteristics that can be altered or damaged.

Orbital debris, often called space junk, includes defunct satellites, spent rocket stages, and fragments created by collisions or explosions. Even small pieces can travel at speeds capable of destroying operational spacecraft.

Kessler Syndrome describes a cascade effect in which collisions generate debris that causes further collisions, potentially rendering entire orbital regions unusable. This concept, proposed in the late twentieth century, is increasingly relevant as launch rates rise.

Planetary protection encompasses efforts to prevent biological contamination between Earth and other worlds. Forward contamination involves carrying Earth life to other planets, while backward contamination involves bringing potential extraterrestrial organisms to Earth. Both pose scientific and ethical risks.

Mega constellations are large networks of satellites operated by single entities, often numbering in the thousands or tens of thousands. They raise concerns related to debris, astronomy, and equitable access to orbital space.

In situ resource utilization refers to the use of local materials, such as lunar ice or Martian soil, to support space activities. While potentially essential for sustainability, it also raises questions about exploitation and environmental impact.

Terraforming describes the deliberate alteration of another planet's environment to make it more Earth like. It is one of the most ethically charged proposals in space exploration, particularly when weighed against the possibility of preserving existing environments.

The seven generations principle, drawn from indigenous governance traditions, emphasizes considering the impacts of decisions on many future generations. It offers a powerful contrast to short term decision making.

Finally, this book uses the framework of two timelines to explore possible futures. One represents an extraction driven path that leads to environmental degradation. The other represents a stewardship driven path that leads to sustainable exploration. The reality, as we will see, is likely to fall somewhere in between, shaped by ongoing struggle and choice.

A road map for the journey ahead

The chapters that follow are organized to build a comprehensive picture of the challenge and the opportunity before us. The book begins with a look at Earth's environmental crises and the lessons they offer for space. It then moves outward to examine specific environments, including Mars, the Moon, asteroids, and ocean worlds, highlighting what is at stake in each case.

Next, the book explores the threats posed by unregulated commercial activity, debris accumulation, and weak governance. It then turns to ethical questions, including whether planets can have rights and how colonial history informs space exploration. From there, it examines practical pathways toward sustainability, including technological innovations and governance models.

The final sections present two possible futures in vivid detail and introduce the growing movement working to steer humanity toward the more responsible path. The book concludes by returning to the central question of what kind of spacefaring species we choose to become.

An invitation

This introduction has laid out an argument, but it is also an invitation. The chapters ahead will ask you to think beyond familiar frames and to consider space not as a distant abstraction, but as an environment already shaped by human choices. They will challenge assumptions about inevitability and progress, and they will ask uncomfortable questions about responsibility and restraint.

At the same time, they will offer reasons for cautious hope. Awareness is growing. Solutions are emerging. People across disciplines and cultures are beginning to insist that space can and must be treated differently.

The journey begins on Earth, with the lessons we learned too late. It then carries us outward, to worlds still largely untouched, where those lessons can still make a difference.

We are at a cosmic turning point. The path ahead is not fixed.

Turn the page. The future is waiting.

1

The Day We Broke Earth

June 7, 2023

Sarah woke before her alarm because something felt off in the room.

The light pressing through the blinds did not have the clarity or softness she associated with early summer mornings in New York. It was dim and thick, tinged with a deep amber hue that reminded her of photographs taken decades ago, when film degraded and colors bled together. She lay still for a moment, listening to the low hum of the city through her closed windows, trying to decide whether she had overslept or whether a storm was rolling in.

When she finally rose and crossed her apartment, the air felt heavy against her skin, as if the room had been sealed for days. She lifted the blinds slowly, already uneasy in a way she could not name. Outside, the city had been transformed. The sky was not blue, nor gray, nor even storm darkened. It was orange, a dense and unnatural orange that flattened the skyline into shadow. Buildings faded into one another, their outlines softened as if erased with a thumb. The far end of her block disappeared entirely into haze.

Her phone vibrated on the counter behind her. Air quality alert. Hazardous conditions. Limit time outdoors. Schools closed.

Sarah stared out the window, trying to reconcile what she was seeing with what she knew. She was thirty four years old, a science teacher at a public middle school, and she had spent years explaining climate systems to children. She had taught them how rising temperatures dried forests, how drought turned ecosystems into tinder, how wind carried smoke across borders that existed only on maps. She had shown them graphs of increasing wildfire acreage and photographs of charred landscapes in California and Australia.

None of those lessons prepared her for the smell creeping into her apartment. It seeped through the window seals and vents, sharp and bitter, catching in her throat. Her eyes burned. The smoke carried the weight of distance and scale, forests burning hundreds of miles away in Canada, their ash now settling over one of the largest cities in the world.

For the first time, climate change was not abstract. It was not an article she assigned her students or a unit she taught in the spring. It was the sky outside her window, altered so completely that it no longer felt like Earth at all.

The Visible Collapse

The orange skies over New York City were not an isolated event. They were the visible intersection of long building forces that had been measured, modeled, and warned about for decades. By the early 2020s, Earth's climate system had begun to display its stress openly, in ways that could no longer be dismissed as statistical noise or rare coincidence.

According to data compiled by NASA and assessed by the Intergovernmental Panel on Climate Change, Earth's average surface temperature had increased by approximately 1.2 degrees Celsius above pre industrial levels by 2023. This number often failed to register emotionally, partly because it sounded modest and partly because it was framed as a global average. Yet in geological and biological terms, it represented an extraordinary shift compressed into an alarmingly short period of time. The last transition from an ice age into a stable inter glacial climate unfolded over thousands of years. Human driven warming achieved a significant fraction of that change in little more than a century.

The concentration of carbon dioxide in Earth's atmosphere told an even clearer story. For at least eight hundred thousand years prior to industrialization, atmospheric CO_2 levels fluctuated between roughly 180 and 280 parts per million, rising and falling slowly with natural climate cycles. By 2023, those levels exceeded 425 parts per million, a concentration unseen in human evolutionary history. This increase was not subtle or gradual when viewed on geological timescales. It was a spike, sharp and unmistakable, driven almost entirely by the combustion of fossil fuels and the destruction of carbon absorbing ecosystems.

As temperatures rose, the climate system responded not smoothly, but unevenly, amplifying extremes. Heat waves grew longer and more intense. Rainfall patterns shifted, concentrating precipitation into fewer, heavier events while leaving long dry intervals between them. The atmosphere, warmed by additional energy, held more moisture, turning storms into more destructive forces when conditions aligned.

In the summer of 2021, the Pacific Northwest experienced a heat dome that stunned scientists and residents alike. Temperatures in parts of British Columbia reached 120 degrees Fahrenheit, a figure that would have been considered physically implausible for the region only a decade earlier. Hundreds of people died from heat related causes. Entire towns, such as Litton, were destroyed by fires ignited under unprecedented conditions. Climate scientists later concluded that such an event would have been virtually impossible without human driven climate change.

The following year, in 2022, Pakistan endured catastrophic flooding after record monsoon rains combined with accelerated glacial melt. At the peak of the disaster, roughly one third of the country lay underwater. Millions were displaced. Crops were destroyed. Infrastructure collapsed. The floods did not simply cause immediate devastation. They undermined food security, increased disease risk, and locked vulnerable populations into cycles of recovery and loss that would persist for years.

By 2023, wildfires had become a global phenomenon rather than a regional one. Southern Europe experienced intense fires across the Mediterranean basin, consuming forests and threatening ancient towns during prolonged heat waves. In Canada, fires burned with such scale and intensity that smoke crossed continents, dimming skies in the northeastern United States and triggering public health emergencies in cities unaccustomed to such conditions. The smoke that Sarah saw outside her window was part of a plume that stretched thousands of kilometers, a visible reminder that atmospheric systems did not respect political boundaries.

The trend continued into 2024. Record breaking heat gripped South America, straining electrical grids and water supplies. Parts of Southeast Asia experienced deadly combinations of heat and humidity that pushed human tolerance limits. In the Atlantic, ocean temperatures surged to levels that fueled unusually powerful storms, while marine heat waves stressed coral ecosystems already weakened by decades of warming.

Alongside these dramatic events ran a quieter but equally devastating crisis. Earth's biodiversity was collapsing at a pace unmatched in recorded history. A landmark study led by biologist Gerardo Ceballos described the phenomenon as "biological annihilation," noting that approximately sixty percent of wildlife populations had declined since 1970. This loss was not limited to charismatic species or remote habitats. It extended across mammals, birds, reptiles, amphibians, and fish, reshaping ecosystems from the ground up.

Insects, often overlooked, were among the hardest hit. Long term studies in Germany revealed declines of up to seventy five percent in flying insect biomass over just a few decades. The implications were profound. Insects pollinated crops, decomposed waste, and formed the foundation of countless food webs. Their disappearance destabilized systems that humans depended on, often without realizing it.

The oceans, which absorbed more than ninety percent of the excess heat trapped by greenhouse gases, underwent changes that were both vast and subtle. As carbon dioxide dissolved into seawater, it formed carbonic acid, lowering the ocean's pH in a process known as acidification. This chemical shift made it harder for organisms such as corals, shellfish, and plankton to build calcium carbonate structures. Over time, reefs weakened and died. By the early 2020s, scientists estimated that roughly half of the Great Barrier Reef had been lost to repeated bleaching events driven by marine heat waves.

Ice sheets provided another unmistakable signal. Greenland and Antarctica began losing mass at accelerating rates, contributing to global sea level rise. Satellite measurements showed that Greenland alone was shedding hundreds of billions of tons of ice per year. Melt water lubricated glaciers, hastening their flow into the ocean. These changes were not easily reversible. Once initiated, they carried momentum that would continue for centuries.

For Sarah, these facts arrived piecemeal through news articles, lesson plans, and late night reading. They felt overwhelming in isolation, but together they formed a pattern that was impossible to ignore. The planet was changing rapidly, and the changes were no longer confined to distant places or future projections. They were unfolding in real time, visible in the air she breathed and the sky she saw each morning.

The Invisible Damage

While fires darkened skies and floods erased neighborhoods, another layer of environmental damage accumulated quietly, largely unseen. These were the impacts that did not announce themselves with dramatic imagery, but which penetrated bodies, soils, and systems with equal persistence. They were harder to photograph and easier to ignore, yet no less consequential.

Microplastics exemplified this hidden crisis. Once thought to be confined to oceans and landfills, microscopic fragments of plastic were discovered throughout the environment. By 2022, studies had detected microplastics in human bloodstreams, confirming that these particles were not merely external pollutants but internalized ones. Researchers found them in clouds and rainwater, carried aloft by atmospheric currents and deposited far from their sources. They were retrieved from the deepest ocean trenches, embedded in sediments kilometers below the surface. Perhaps most unsettling, they were found in human placentas, crossing biological barriers once assumed to be protective.

Plastics did not biodegrade in any meaningful sense. They fragmented, breaking into ever smaller pieces that persisted for centuries. Their chemical additives, designed to provide flexibility, durability, or resistance to heat, leached into surrounding tissues. The long term health implications were only beginning to be understood, but early evidence suggested links to inflammation, endocrine disruption, and metabolic disorders.

Alongside plastics were per and polyfluoroalkyl substances, collectively known as PFAS. These synthetic chemicals were prized for their resistance to heat, water, and oil, properties that made them useful in everything from nonstick cookware to firefighting foam. Their molecular structure, however, rendered them extraordinarily persistent. Once released, they did not break down under natural conditions, earning them the name “forever chemicals.”

PFAS contamination was detected in drinking water supplies across the globe, as well as in food packaging, soil, and wildlife. Studies linked exposure to increased risks of certain cancers, immune system suppression, and developmental issues. Like microplastics, PFAS illustrated a broader pattern in industrial society: chemicals were introduced widely before their consequences were fully understood, and once their dangers became clear, they proved difficult or impossible to remove.

Soil, often treated as inert substrate, was undergoing its own crisis. Industrial agriculture prioritized short term yields through intensive tilling, monocropping, and heavy chemical inputs. These practices degraded soil structure, reduced organic matter, and accelerated erosion. The United Nations estimated that at current rates of loss, humanity had roughly sixty years of productive topsoil remaining. This figure was not a prediction of total depletion, but a warning that the foundation of global food systems was eroding faster than it could be replenished.

Aquifers told a similar story beneath the surface. The Ogallala Aquifer, which supported much of the agricultural output of the central United States, was being depleted far faster than it could recharge. In parts of India, groundwater extraction exceeded sustainable limits by wide margins, driven by growing populations and water intensive crops. As water tables dropped, wells ran dry, forcing farmers to drill deeper at greater cost or abandon land altogether. These invisible changes threatened food security long before shortages appeared on grocery shelves.

For Sarah, these issues felt harder to grasp than fires or floods. They lacked a single moment of impact, no clear image to anchor concern. Yet as she learned more, she began to understand that they represented a deeper kind of damage, one that seeped into systems quietly until resilience was gone.

She began to notice how often environmental harm occurred out of sight, buffered by distance, scale, or complexity. The planet was not only warming.

It was being altered chemically, biologically, and structurally in ways that undermined its capacity to support life as humans had known it.

What made the invisible damage so dangerous was not only its scale, but its intimacy. These were not distant harms affecting polar bears or remote islands. They were entering human bodies, altering biological processes at levels too small to feel yet too large to dismiss. Unlike a flood or a fire, there was no moment when alarms sounded and evacuation orders were issued. The damage accumulated quietly, often noticed only when systems failed.

Scientists struggled to communicate this kind of risk because it resisted narrative simplicity. Microplastics did not arrive all at once. PFAS exposure did not produce immediate illness in most cases. Soil degradation did not announce itself with a single catastrophic harvest failure. Instead, these forces eroded resilience over time, reducing margins of safety until ordinary stresses became crises.

Soil offered a useful analogy. Healthy soil functioned as a living system, rich with microbes, fungi, and organic matter that stored carbon, retained water, and supported plant growth. Industrial farming treated soil as a medium to hold crops upright while fertilizers delivered nutrients directly. Over time, this approach stripped soil of life, compacted its structure, and left it vulnerable to erosion. When droughts came, degraded soil could not retain moisture. When rains arrived, it washed away. The loss was incremental, often invisible until fields failed entirely.

Aquifers behaved similarly. Groundwater systems formed over thousands of years as rain slowly filtered downward, accumulating in porous rock and sediment. Industrial agriculture and urban growth extracted water far faster than natural recharge could replace it. The Ogallala Aquifer became emblematic of this imbalance. Supporting nearly one fifth of United States agricultural production, it was being drained unevenly, with some regions projected to lose practical access within decades. India faced parallel challenges, with millions of wells drawing from shared underground reserves without coordinated management.

For Sarah, these realizations deepened her unease. She began to see environmental collapse not as a series of disasters, but as a slow thinning of buffers that once absorbed shocks. The systems humans relied on were losing their elasticity. When stress came, there was less room to bend without breaking.

The invisible damage also exposed a fundamental mismatch between human perception and planetary processes. Humans were attuned to immediate threats, sudden changes, and personal experiences. They struggled with gradual harm, delayed consequences, and collective responsibility. This gap between perception and reality would prove decisive.

The Psychology of Denial

By the time the evidence of environmental collapse became undeniable, the question many people asked was not whether it was happening, but why it had taken so long to respond. The data had been clear for decades. Climate scientists had warned of rising temperatures and cascading impacts since the late twentieth century. Reports grew more confident, not less, as observations matched predictions. Yet meaningful action lagged far behind knowledge.

Part of the answer lay in the structure of shared resources, a problem economists described as the tragedy of the commons. When a resource was shared by many and owned by none, individuals acting in their own short term interest could deplete it, even if everyone understood that doing so harmed the group. The atmosphere functioned as such a commons. Each emitter benefited privately from burning fossil fuels, while the costs were distributed globally and deferred into the future.

Psychological distance compounded the problem. Climate impacts often seemed temporally distant, framed as threats to future generations rather than present ones. They appeared spatially distant, happening in other countries or ecosystems. They felt socially distant, affecting people unlike oneself. These layers of separation reduced emotional urgency, allowing individuals and institutions to postpone action without feeling immediate consequence.

Economic systems reinforced delay. Modern economies were organized around quarterly profits and continuous growth, rewarding decisions that maximized short term returns even when they imposed long term costs. Pollution, habitat destruction, and health impacts were treated as externalities, meaning they were not reflected in the price of goods and services. As long as environmental damage remained unpaid and uncounted, it could be rationalized as acceptable collateral.

Gross domestic product became a proxy for progress despite its blindness to ecological limits. Growth was celebrated even when it depended on liquidating natural capital. The planet, finite by definition, was treated as though it were infinite by assumption. This contradiction sat at the heart of environmental failure.

Cognitive biases further shaped behavior. Optimism bias led people to believe negative outcomes were less likely to affect them personally. Present bias prioritized immediate comfort over future risk. Diffusion of responsibility encouraged passivity by convincing individuals that someone else would act. These tendencies were not moral failures. They were features of human psychology, evolved for survival in small groups facing immediate threats, not for managing global systems over centuries.

Political systems struggled to overcome these forces. Fossil fuel industries invested heavily in lobbying and public messaging to delay regulation and cast doubt on scientific consensus. International coordination faltered as nations prioritized domestic interests over collective outcomes. Agreements such as the Copenhagen Accord and the Paris Agreement represented progress, but their voluntary commitments and enforcement gaps fell short of what scientists said was necessary to stabilize the climate.

Sarah encountered these explanations gradually, through books, documentaries, and conversations with colleagues. She recognized herself in many of them. She recycled, conserved energy, and voted with environmental concerns in mind, yet she also flew to visit family and bought products wrapped in plastic. The system offered few paths to live without contradiction. Understanding this did not absolve responsibility, but it reframed it. The failure was not individual weakness. It was systemic inertia.

The Breaking Point

As the twenty first century progressed, scientists began to describe Earth's condition using a framework known as planetary boundaries. Developed by researchers at the Stockholm Resilience Centre, the framework identified nine critical Earth system processes that regulated the planet's stability, including climate change, biodiversity integrity, land use, freshwater use, and biochemical flows of nitrogen and phosphorus. Crossing these boundaries increased the risk of large scale, irreversible change.

By the early 2020s, evidence suggested that humanity had already crossed at least six of the nine boundaries. Climate change, biodiversity loss, land system change, freshwater use, nutrient cycles, and novel entities such as synthetic chemicals had all exceeded safe operating limits. These transgressions did not guarantee immediate catastrophe, but they increased the likelihood of tipping points, thresholds beyond which systems shifted into new states.

Feedback loops amplified concern. As Arctic ice melted, darker ocean water absorbed more heat, accelerating warming in a process known as Arctic amplification. Thawing permafrost released methane, a potent greenhouse gas, further intensifying warming. The Amazon rain forest, stressed by deforestation and drought, risked crossing a threshold beyond which it could no longer sustain itself as a rain forest, transforming into savanna and releasing vast amounts of stored carbon.

The planet began to resemble a warning system whose alarms were sounding simultaneously. These were not isolated failures, but interconnected signals of a system under strain. The question was no longer whether humanity could avoid all damage, but whether it could prevent cascading collapse.

For Sarah, this realization marked a turning point. One evening, after a long day of teaching and grading papers, she stepped onto her building's roof. The sky had returned to its usual color, but the memory of orange haze lingered. She looked up at the stars, faint but present, and wondered whether humanity could do better elsewhere than it had done here.

The thought unsettled her. If humans could not sustain one planet with abundant water, stable climate, and complex life, what business did they have spreading beyond it?

Looking to the Stars

By the time Sarah began asking those questions, she was not alone. Across the world, people sensed that something fundamental had shifted. Environmental collapse was no longer a distant possibility. It was a lived experience, unevenly distributed but increasingly unavoidable. Fires, floods, heat waves, and invisible toxins all pointed to the same conclusion. The way humanity had organized its relationship with Earth was no longer tenable.

The evidence was overwhelming. The climate was warming faster than ecosystems could adapt. Biodiversity was declining at rates comparable to past mass extinctions. Chemical pollution was altering biological systems in ways still poorly understood. The buffers that once protected societies from disruption were thinning. Earth, resilient but not limitless, was signaling distress.

And yet, even as these realities set in, attention began to shift outward. Rockets lifted off with renewed frequency. Private companies and national agencies spoke of Mars, the Moon, and asteroids as sites of opportunity. Space was framed as a frontier, a backup plan, a place to start over.

The temptation was understandable. Faced with damage at home, humans had always looked elsewhere. But the question lingered, unresolved and urgent. Would humanity carry its wisdom into space, or simply export its habits? Was space a second chance, or a final mistake repeated on a grander scale?

Sarah did not have answers. She only knew that the sky she had seen that morning in June 2023 had changed something inside her. It had collapsed the distance between cause and effect, between abstraction and experience. It had made the stakes unmistakably clear.

We broke Earth. Now we are looking to the stars.

Some believe space is our escape hatch. They are wrong.

In the chapters that follow, we will examine why the future of space is inseparable from the fate of Earth, and how the choices humanity makes beyond this planet will determine whether it becomes a steward of worlds or their undoing.

2

The Great Escape Myth

"I want to die on Mars, just not on impact."

The line landed perfectly. It was confident, humorous, and visionary all at once, a sound bite engineered for an age that rewarded ambition packaged as inevitability. When Elon Musk said it, millions heard not just a joke, but a promise. Humanity would become a space faring species. Mars would be our next home. Earth, damaged and crowded, would no longer be our only option.

This dream inspired engineers, students, investors, and ordinary people searching for hope in a time of environmental anxiety. It suggested continuity in the face of crisis, a future that did not end with sacrifice but with expansion. It implied that technology could outpace consequence, that ingenuity could outrun ecological limits.

It was also dangerously wrong.

The problem was not that humans dreamed of Mars. The problem was what the dream was being asked to do. It was being framed not as exploration, or even as survival insurance, but as escape. A backup planet. A release valve for a civilization unwilling to confront the systems that were unraveling its only proven home.

Understanding why this idea took hold requires sympathy, not scorn. The fantasy did not arise from ignorance. It arose from fear, exhaustion, and the deeply human desire to believe there is always somewhere else to go.

The Seductive Fantasy

The notion that humanity's destiny lay among the stars had deep roots. Long before rockets existed, people looked up and imagined meaning written across the night sky. Exploration narratives had always carried moral weight, casting movement outward as progress itself. In this framing, stagnation was failure, and expansion was virtue.

Space amplified this instinct. It offered a canvas vast enough to absorb projection. Against the backdrop of climate disruption, political paralysis, and ecological loss, the idea of starting over somewhere clean and empty felt emotionally irresistible. It spoke directly to the survival instinct, not just of individuals, but of the species. If Earth was fragile, then diversification seemed prudent. Do not put all your eggs in one planetary basket.

Pioneer mythology reinforced the appeal. In American cultural memory, the frontier symbolized renewal through movement. When land became crowded or resources scarce, settlers moved west. Hardship became opportunity. Reinvention followed relocation. This narrative was deeply flawed in its erasure of indigenous peoples and ecological damage, but it remained powerful. Mars became the new frontier, and space advocates consciously or unconsciously borrowed its language. Colonists would be self reliant. Risks would forge character. Sacrifice would be noble.

Technological optimism filled in the gaps. The same era that produced environmental breakdown also produced astonishing advances in computing, biotechnology, and materials science. If humans could build reusable rockets, sequence genomes, and simulate entire ecosystems, surely they could engineer life support on another planet. Space was framed as an engineering challenge rather than an ecological one, and engineering challenges felt solvable.

Prominent figures articulated these visions with confidence. Elon Musk spoke openly of a self sustaining city on Mars with a population of one million people by mid century. Jeff Bezos imagined billions of humans living in massive orbital habitats, freeing Earth to recover as a preserved garden. Robert Zubrin argued passionately that Mars was not just reachable, but necessary, laying out detailed plans for in situ resource utilization and permanent settlement.

These visions resonated because they offered hope without requiring fundamental change. They suggested that humanity could continue its existing trajectory, powered by markets and innovation, while sidestepping the need to radically rethink consumption, growth, and inequality. Instead of asking how to live within limits, they asked how fast limits could be outrun.

Popular culture reinforced the story. Films depicted Mars as harsh but conquerable, a place where grit and intelligence prevailed. Television series imagined complex, inhabited solar systems where humanity's presence felt inevitable. Novels explored detailed futures in which planetary engineering transformed hostile worlds into habitable ones. These stories did not invent the escape myth, but they normalized it, embedding it into the cultural imagination.

Alex absorbed all of this quietly. At twenty eight, working as a software engineer, he followed space news obsessively. He watched launch streams late into the night and read mission updates with the focus others reserved for sports scores. He had opened a separate savings account labeled simply "Mars," adding small amounts whenever he could, half joking, half sincere.

The idea that he might someday leave Earth felt less like abandonment and more like participation in something historic.

When Earth news felt bleak, space news felt clean. Rockets lifted upward. Problems seemed bounded. The fantasy did not deny Earth's crisis. It simply offered a door marked exit.

The Math Does Not Work

The escape narrative unraveled when numbers replaced metaphors.

At present, sending mass to Mars remained extraordinarily expensive. Even optimistic estimates placed the cost at well over one hundred thousand dollars per kilogram, accounting for launch, transfer, landing, and life support infrastructure. A single human, when accounting for body mass, food, water, oxygen, shelter, equipment, and redundancy, required at minimum roughly one thousand kilograms to survive for any meaningful duration. This placed the baseline cost per person well above one hundred million dollars, even before accounting for return capability or long term sustainability.

These figures were not pessimistic. They were generous.

Even if costs dropped by a factor of one thousand, an improvement far beyond any historical precedent in transportation economics, the price would still place Mars out of reach for all but a tiny fraction of humanity. With a global population exceeding eight billion people, the idea of relocating even one percent would require resources that dwarfed current global economic output.

Energy requirements posed an even more fundamental constraint. Reaching Mars required achieving a change in velocity on the order of five to seven kilometers per second, depending on trajectory. The kinetic energy needed to accelerate a kilogram of mass to that speed was immense. When multiplied by the mass of life support systems and shielding required to protect humans from radiation, the total energy demand became staggering.

To move even a million people to Mars, each with minimal support, would require energy comparable to a significant fraction of total annual global energy production. This comparison assumed perfect efficiency, which did not exist. Real systems incurred losses at every stage. Rockets burned fuel to lift fuel. Energy compounded into waste heat and exhaust.

Timelines further undermined the fantasy. Even the most optimistic projections envisioned a few hundred or thousand people living on Mars by mid century. During that same period, Earth's population would grow by billions. Climate impacts would intensify within decades, not centuries. A backup that arrived too late and served too few was not a backup. It was a distraction.

Life support challenges sharpened the contrast. Mars possessed less than one percent of Earth's atmospheric pressure, lacked a global magnetic field to shield against radiation, and contained soil toxic to human biology. Every breath, every sip of water, every calorie would need to be engineered and maintained artificially. A single systemic failure could mean total loss of life.

Earth, even damaged, remained astonishingly forgiving by comparison. Its atmosphere regenerated oxygen. Its ecosystems recycled waste. Its magnetic field deflected cosmic radiation. Its gravity supported long term human health. By any rational metric, Earth was millions of times more habitable than Mars, even in crisis.

Alex encountered these calculations gradually, first as curiosities, then as constraints, and finally as revelations. He did not abandon his fascination with space. He abandoned the idea that space could replace Earth. The numbers were not ideological. They were physical.

Escape, he realized, was not just unlikely. It was impossible at any scale that mattered.

You Can't Escape Yourself

History offered a sobering lens through which to view the promise of a fresh start. Again and again, humans had believed that moving to new lands would free them from the failures of old ones. European colonization of the Americas was framed as renewal, opportunity, and rebirth. What followed was not reinvention, but replication. The same systems of extraction, hierarchy, and exploitation were carried across oceans and imposed on new ecosystems and peoples. Forests were cleared. Rivers were redirected. Indigenous societies were destroyed. The problems that drove expansion were not solved. They were exported and magnified.

This pattern was not incidental. It emerged from the continuity of human behavior and institutions. People did not arrive as blank slates. They arrived with economic models, power structures, and assumptions about entitlement. New environments did not erase old incentives. They merely delayed consequences.

Psychological research reinforced this lesson. Isolated and extreme environments tended to amplify human conflict rather than transcend it. The Biosphere 2 experiment in the early 1990s was intended to demonstrate the feasibility of closed ecological systems. Instead, it became a case study in social fracture. Crew members split into factions. Food shortages worsened tensions. Oxygen levels dropped unexpectedly due to soil chemistry interactions that designers had not fully anticipated. The system failed not because of a lack of intelligence, but because complexity overwhelmed control.

Studies of Antarctic research stations revealed similar dynamics. Small groups in confined, high stress environments experienced elevated rates of depression, interpersonal conflict, and cognitive decline. Submarine crews showed comparable patterns despite rigorous selection and training. The Mars500 simulation, which isolated volunteers for over a year to mimic a Mars mission, documented mood disturbances, social withdrawal, and circadian disruption. None of these outcomes were moral failures. They were predictable human responses to confinement, monotony, and perceived risk.

These findings mattered because Mars would not be a frontier of freedom. It would be the most controlled environment humans had ever inhabited. Every system would be closed. Every error would compound. Every conflict would unfold without escape. The romance of independence dissolved under the reality of total dependence on machines and each other.

Underlying these challenges was a deeper systems problem. The economic model that had driven Earth's environmental collapse was not planet specific. Capitalism as practiced required continuous growth, expanding markets, and increasing resource throughput. On a finite planet, this logic led to overshoot. On a smaller and harsher planet like Mars, it would lead to failure even faster.

As Albert Einstein famously observed, problems could not be solved using the same thinking that created them. Transporting an extractive, growth driven system to another world would not produce sustainability. It would reproduce collapse under tighter constraints.

The question of who would go further exposed the moral dimensions of escape. Space colonization, at least in its early stages, would be accessible only to the wealthy, the highly skilled, or those selected by institutions with power. This was not speculation. It was already true of spaceflight. The result would be a new stratification of humanity, divided between those who could leave and those who could not.

This division carried ethical weight. Abandoning Earth would mean abandoning the vast majority of humanity. It would frame survival as a privilege rather than a shared responsibility. Ethicists warned that such a future risked normalizing inequality on a planetary scale, with profound consequences for how humans valued one another.

Alex felt this tension acutely. As someone with the education and income that might someday qualify him for off world opportunities, he recognized the temptation to prioritize personal escape. Yet the more he considered it, the more hollow it felt. Leaving Earth would not absolve him of responsibility. It would merely relocate it.

The Real Reason to Go

Rejecting escape did not require rejecting space. In fact, some of the strongest arguments for space exploration emerged precisely because Earth was fragile.

Space exploration had historically expanded human perspective. Astronauts returning from orbit spoke consistently of the overview effect, a cognitive shift that came from seeing Earth as a single, border less sphere suspended in darkness. Chris Hadfield described this perspective as trans formative, fostering a sense of unity and responsibility difficult to achieve from the ground. Images of Earth from space had fueled environmental movements and reshaped cultural understanding of planetary limits.

Scientific discovery provided another compelling rationale. Studying other planets deepened understanding of Earth itself. Comparative planetology revealed how atmospheres evolved, how climates stabilized or destabilized, and how life might arise or fail. Searching for life beyond Earth informed biology, chemistry, and philosophy alike. It contextualized humanity's place in the universe.

There were also legitimate arguments related to long term species survival. Earth had experienced mass extinctions before. Asteroid impacts, while rare, were real risks. A multi planetary species would be more robust over geological timescales. However, this logic held only if expansion was deliberate, sustainable, and complementary to Earth stewardship. A broken Earth paired with a fragile Mars colony did not constitute resilience. It constituted dispersion of vulnerability.

The crucial distinction lay in framing. Space could be a complement to Earth, not a replacement. Exploration could coexist with restoration. Innovation could serve both off world and on world sustainability. Closed loop life support systems developed for space could inform circular economies on Earth. Energy technologies refined for remote habitats could accelerate decarbonization at home.

This was not an either or proposition. It was a both and responsibility. Fix Earth while exploring space. Learn from one to protect the other. Use the harshness of space as a teacher rather than an escape route.

Alex's enthusiasm did not fade. It matured. He still watched launches, but he also volunteered with local environmental groups. He still dreamed of Mars, but he no longer imagined it as a refuge from Earth. He imagined it as a test of whether humanity could finally live within limits.

Space Is Not Plan B

The great escape myth persisted because it offered comfort. It promised continuity without reckoning. It suggested that consequences could be outrun rather than addressed. But comfort was not truth.

The numbers did not support escape. The physics did not allow it. The psychology warned against it. History cautioned that relocation without transformation led only to repetition.

Earth was not disposable. It was not replaceable. It was not optional.

Space, when stripped of fantasy, revealed itself not as an exit, but as a mirror. How humans behaved beyond Earth would reflect who they were, not who they wished to be. If they carried exploitation outward, they would export collapse. If they carried stewardship, they might earn expansion.

The future would not be decided by whether humans reached Mars. It would be decided by how they did so, and whether they learned from the planet they had already pushed to its limits.

If we cannot escape Earth, and we are going to space anyway, then we must ask a harder question.

What does Earth teach us about how not to treat other worlds?

3

What Earth Teaches Us

Astronauts struggle to describe the moment when Earth first comes fully into view. Words feel insufficient for the scale of perception involved, yet across decades of spaceflight, their descriptions converge with remarkable consistency. From orbit, political boundaries vanish. National identities dissolve. The planet appears not as a collection of territories, but as a single, luminous system suspended in darkness.

Frank White, who coined the term "overview effect" after interviewing dozens of astronauts, described it as a cognitive shift in awareness, a sudden understanding of Earth as an interconnected whole. Edgar Mitchell, returning from the Apollo 14 mission, spoke of experiencing an overwhelming sense of unity and responsibility. Ron Garan, who spent months aboard the International Space Station, later reflected that from space you could not see borders or wars, only a fragile organism powered by a thin atmosphere clinging to a rocky sphere.

This perspective mattered because it revealed a truth often obscured at ground level. Earth was not merely the stage on which human activity unfolded. It was the system that made all activity possible. Every breath, every harvest, every civilization depended on processes so vast and continuous that they faded into background assumption. Seeing the planet whole stripped away that invisibility.

For those who experienced it, the overview effect was not abstract inspiration. It was an ethical confrontation. If Earth was a single system, then harm in one place resonated everywhere. If there was no visible boundary between nations, then responsibility could not be neatly contained. Space, paradoxically, did not pull astronauts away from Earth. It pulled them toward it, with renewed clarity.

This chapter begins from that vantage point, not to romanticize the planet, but to take it seriously as a teacher. If humanity intended to live beyond Earth, then Earth itself offered the only working model of how life endured over time.

Earth as Our Only Working Model

Earth remained unique not because it was perfect, but because it worked. It was the only known example of a planet that had sustained complex life for billions of years. This endurance was not accidental. It emerged from a network of self regulating systems refined through evolutionary time on a scale far beyond human experience.

Life appeared on Earth roughly 3.8 billion years ago, almost as soon as conditions allowed it. Since then, the planet had endured asteroid impacts, ice ages, volcanic upheavals, and massive shifts in atmospheric composition. Through it all, life persisted, adapting and reshaping the environment in feedback loops that stabilized conditions within narrow bounds. This was not balance in a static sense, but balance through dynamic change.

The Gaia hypothesis, proposed by James Lovelock and further developed with Lynn Margulis, suggested that life and Earth's physical systems co evolved in ways that maintained habitable conditions. The idea was often misunderstood as mystical or teleological, but at its core it described feedback mechanisms grounded in chemistry, biology, and physics. Photosynthetic organisms altered atmospheric composition, producing oxygen that enabled more complex life. Microbes influenced cloud formation. Vegetation affected albedo and rainfall patterns. No single organism controlled the system, yet the system exhibited emergent regulation.

Earth functioned as a closed loop with respect to matter and an open system with respect to energy. Solar radiation provided a constant input of energy, driving weather, photosynthesis, and ecological productivity. Matter, by contrast, cycled endlessly. Carbon moved through the atmosphere, oceans, rocks, and living organisms. Nitrogen was fixed by bacteria, incorporated into life, released, and reused. Water evaporated, condensed, precipitated, and flowed back to the sea. Waste did not accumulate indefinitely. It became input for something else.

Biodiversity played a central role in this resilience. Diverse ecosystems contained redundancy, multiple species capable of fulfilling similar roles. When one failed, others compensated. This diversity buffered shocks and prevented localized disturbances from cascading into systemic collapse. The planet's stability was not the product of efficiency, but of richness and overlap.

These characteristics mattered profoundly for space. Closed ecological systems were not theoretical. Earth proved they were possible. Yet Earth also demonstrated how complex they were. The planet's life support systems were not engineered in decades. They emerged over billions of years of evolution, punctuated by mass extinctions and long recovery periods. Attempting to replicate even a fraction of this complexity inside a space habitat or a Martian base was an extraordinary challenge.

Isabella confronted this reality daily. As an environmental scientist working on closed loop life support systems for future Mars missions, she studied

Earth obsessively. She mapped nutrient flows in forests, analyzed microbial interactions in soil, and modeled atmospheric exchanges between plants and air. The more she learned, the more humbled she became. Systems that appeared simple from a distance revealed layers of interdependence that defied reduction.

She once summarized the problem to a colleague with quiet disbelief. We are trying to build in ten years what Earth took three billion years to refine. The statement was not defeatist. It was diagnostic. It identified the scale of the task and the necessity of learning from Earth rather than attempting to dominate it.

Bio mimicry offered one promising approach. Advocated by thinkers such as Janine Benyus, bio mimicry treated nature not as a resource to be exploited, but as a library of solutions tested by time. Engineers studied how leaves captured sunlight to design more efficient solar cells. Designers mimicked termite mounds to regulate temperature in buildings. Materials scientists learned from spider silk and seashells to create strong, lightweight composites.

In the context of space habitats, bio mimicry suggested designs that prioritized circular flows, redundancy, and adaptability. Rather than forcing rigid control, systems could be built to self adjust within defined parameters, emulating ecological processes. Earth did not eliminate variability. It absorbed it.

Understanding Earth as a system was not an exercise in reverence alone. It was practical instruction. Any attempt to live beyond Earth without internalizing these lessons would be fragile by definition.

The Lessons We Ignored

Despite living within Earth's systems, humans repeatedly failed to learn from them. The pattern was not ignorance, but selective attention. Short term gains overshadowed long term costs. Control was favored over adaptation. Efficiency displaced resilience.

The Industrial Revolution exemplified this dynamic. It promised unprecedented prosperity through mechanization and fossil energy, and in many respects it delivered. Life expectancy increased. Productivity soared. Entire societies transformed. Yet these benefits came with costs that were initially externalized and largely ignored. Air pollution darkened cities and poisoned lungs. Rivers became industrial sewers. Coal mining devastated landscapes and communities. The economic gains were unevenly distributed, while the environmental consequences accumulated silently.

The lesson was clear in retrospect. Externalized costs did not disappear. They returned as health crises, ecological damage, and social instability. Industrialization without accounting for environmental limits created debts that later generations were forced to pay. In space, the temptation to repeat this pattern was obvious. Rapid development, driven by competition and profit, could prioritize short term extraction over long term viability.

Colonial extraction economies revealed a similar structure at a larger scale. Empires extracted rubber from the Congo, silver from Potosí, timber from the Pacific Northwest, and countless other resources from colonized lands. The wealth flowed outward. The damage remained. Forests were stripped. Soils eroded. Indigenous communities were displaced or destroyed. These systems were efficient for accumulation, but catastrophic for ecosystems and human lives.

The logic underlying asteroid mining or lunar resource extraction bore uncomfortable similarities. Remove resources where governance was weak or nonexistent. Justify harm as progress. Assume the frontier absorbed damage without consequence. Earth demonstrated the fallacy of this thinking repeatedly. There was no extraction without impact. There was no external elsewhere.

The Green Revolution of the mid twentieth century offered a more complex lesson. New crop varieties, fertilizers, and irrigation dramatically increased food production, averting famine for millions. Yet the technological success masked deeper costs. Intensive farming depleted soils, drained aquifers, polluted waterways with nitrogen runoff, and reduced genetic diversity in crops. Farmers became dependent on external inputs and vulnerable to market shocks. In some regions, debt and ecological stress contributed to rising suicide rates among agricultural workers.

The intention had been noble. The outcome was mixed. The lesson was not that technology failed, but that technological fixes introduced new dependencies and risks when divorced from ecological context. Proposals to terraform planets or engineer life for space carried similar dangers. Complex systems responded in nonlinear ways.

Nuclear waste offered perhaps the starkest warning. Nuclear energy was developed before long term waste storage was solved. Highly radioactive materials with lifespans measured in tens of thousands of years accumulated without a clear plan for stewardship. Facilities like Yucca Mountain became symbols of political stalemate and ethical deferral. Humanity created a problem that outlasted institutions, languages, and civilizations.

Space debris mirrored this challenge in orbit. Satellites and fragments accumulated, increasing collision risks and threatening access to space for future generations. The pattern was familiar. Act first. Manage later. Leave consequences behind.

These cases were not indictments of progress. They were patterns. They revealed what happened when systems were designed around short horizons and narrow incentives. Earth recorded these lessons patiently, even as humans failed to read them.

Nuclear waste crystallized a deeper failure of imagination. When the first reactors were built, engineers understood that radioactive byproducts would remain dangerous for periods far exceeding any political system's planning horizon. Yet development proceeded regardless, justified by urgency and optimism that future solutions would emerge. Decades later, those solutions remained incomplete. Waste was stored temporarily, moved repeatedly, and debated endlessly. The problem was not technical alone. It was ethical. Decisions made for immediate benefit imposed obligations on people thousands of years into the future, people who had no voice in the choice.

Space debris followed the same trajectory on a compressed timescale. Early spaceflight treated orbit as empty and forgiving. Satellites were launched without end of life plans. Explosions and collisions multiplied fragments. Today, orbital debris threatens critical infrastructure and human life in space. The lesson repeated itself with unsettling speed. When environments appeared vast or empty, humans behaved as though consequences would dissipate. Earth had already shown this assumption to be false.

These patterns mattered because they revealed a consistent blind spot. Humans excelled at solving immediate problems within narrow frames. They struggled with cumulative effects, long duration, and shared responsibility across generations. Earth did not hide these lessons. It displayed them repeatedly. What was lacking was not evidence, but integration.

Isabella encountered this truth not in theory, but in failure analysis. Every closed loop system she studied contained points of fragility. Nutrient imbalances cascaded. Microbial populations shifted unexpectedly. Redundancy proved insufficient without diversity. Each malfunction echoed something Earth had already experienced on a planetary scale. The difference was that

Earth had time to recover. A space habitat would not.

The lesson was not that humans should abandon ambition. It was that ambition without humility produced systems that broke under their own assumptions.

Indigenous Wisdom

Long before industrial science articulated systems thinking, many Indigenous cultures lived within it. These knowledge systems were not static traditions preserved unchanged across millennia. They were dynamic, empirical, and adaptive, grounded in close observation of local ecosystems and refined through lived experience over generations.

The Haudenosaunee Confederacy articulated one of the clearest expressions of long term responsibility through the Seven Generations principle. Decisions were evaluated based on their impact seven generations into the future, roughly one hundred and forty years. This time frame altered priorities fundamentally. Short term gains that produced long term harm were rejected. Stewardship became a moral obligation rather than a policy preference.

Applied to space, this principle posed unsettling questions. What would a Martian settlement look like not in its first decade, but in the year 2165. Would its systems regenerate resources or merely consume them more slowly. Would its waste accumulate or cycle. Would its culture reinforce care or extraction. Western decision making rarely extended this far because political and economic incentives discouraged it. The result was a chronic inability to plan for durability.

Aboriginal Australian land stewardship provided another instructive model. For over sixty five thousand years, Indigenous Australians managed landscapes through practices such as controlled burning, seasonal movement, and species specific harvesting. These practices shaped ecosystems so effectively that European settlers mistook them for untouched wilderness. In reality, they were deeply managed systems designed to enhance resilience, biodiversity, and productivity.

The concept of Country in Aboriginal thought did not separate land from people. Country was a living relationship encompassing soil, water, animals, ancestors, and responsibility. Damage to land was damage to community. Knowledge was encoded in stories, ceremonies, and practical skills, transmitted across generations. This was not mysticism. It was long term environmental management optimized through observation and feedback.

Māori concepts of kaitiakitanga similarly framed humans as guardians rather than owners. Stewardship implied obligation, reciprocity, and humility. The legal recognition of the Whanganui River as a person reflected this worldview, acknowledging that ecosystems possessed intrinsic value beyond utility. Humans acted as caretakers, accountable to past and future generations.

These frameworks challenged dominant Western assumptions. Capitalism prioritized quarterly earnings. Democratic systems operated on election cycles measured in years. Individual lifespans biased decision making toward immediacy. Judeo Christian interpretations emphasizing dominion over Earth reinforced hierarchies that placed humans above nature rather than within it. Enlightenment philosophy further separated mind from matter, culture from environment.

The result was a civilization structurally ill equipped for long term stewardship. Indigenous knowledge systems did not offer a blueprint to copy wholesale. They emerged from specific places and histories. However, they offered principles applicable beyond their origins. Respect for what is not fully understood. Precaution in the face of irreversible harm. Reciprocity rather than extraction. Acceptance of limits. Recognition that humans are participants in systems, not external managers.

Scholars such as Robin Wall Kimmerer and Kyle Powys Whyte emphasized that these perspectives were not nostalgic alternatives to science, but sophisticated forms of ecological intelligence. They complemented Western methods by extending temporal horizons and re framing values. For space ethics, they provided a missing dimension.

Isabella wrestled with this integration. She did not belong to these cultures, and she knew appropriation was a real risk. Yet she also recognized that ignoring their insights would repeat a familiar arrogance. The challenge was not to adopt symbols, but to internalize principles. Space demanded not just better technology, but better ethics.

The Ultimate Question

The question facing humanity was often framed incorrectly. The issue was not whether humans were technologically capable of going to space. Rockets already flew. Habitats already orbited. The real question was whether humanity was ready.

Readiness was ethical and cultural, not mechanical. It required demonstrating the ability to sustain one planet before attempting to manage another. The analogy was simple. A teenager asking for a car was not judged solely on desire or technical understanding. They were judged on responsibility. Could they care for what they already had. Earth was humanity's bicycle. Space was the car.

By this measure, the evidence was mixed at best. Climate disruption, biodiversity loss, plastic pollution, and chemical contamination suggested that humanity was still learning to live within limits. The verdict was not permanent, but it was clear. Maturity was incomplete.

There was a paradox at the heart of this assessment. Space perspective helped humans appreciate Earth. Seeing the planet whole fostered care. Yet care was required to deserve expansion. Neither could wait for the other. They had to develop together.

The challenge was not to retreat from space, but to rise to it. To let Earth's lessons guide off world behavior. To treat new environments not as blank slates, but as responsibilities.

Earth Is Our Only Teacher

Earth was not humanity's practice planet. It was the only teacher available. Its systems demonstrated what worked, what failed, and what endured. Ignoring those lessons while enrolling in more advanced challenges was a recipe for repetition.

The stakes were high. Space magnified consequences. Errors that Earth could absorb would prove fatal elsewhere. The margin for learning by destruction would vanish.

Yet the story was not one of despair. Earth also showed that sustainability was possible. Life persisted through catastrophe. Systems recovered when allowed time and diversity. Balance emerged not through control, but through relationship.

If humans listened, truly listened, Earth offered guidance. Closed loops mattered. Diversity mattered. Limits mattered. Responsibility across generations mattered.

Understanding what Earth taught us changed the question from whether humanity could go to space to how it should. With that understanding in place, it became possible to examine what was already happening beyond Earth, not in theory, but in practice.

While we debated philosophy and ethics, humans were already on Mars through robots and decisions embedded in code and hardware. Those choices were shaping the future quietly, one mission at a time.

Understanding Earth's lessons, we must now ask what we are actually doing in space, and whether those actions reflect the wisdom we claim to seek.

4

Footprints on Mars (The 2030s)

Commander Jessica Chen's hand trembles as she grips the ladder. Even through the pressurized gloves, she feels the vibration of the spacecraft's structure settling into Martian gravity, lighter than Earth's pull yet heavier than the long months of weightlessness that came before. Through her helmet visor, the rust colored landscape stretches outward toward a pale pink horizon, the ground etched with ancient channels and broken stone. In the two hundred and forty million kilometers behind her, no human has ever stood where she is about to step.

She descends carefully, one rung at a time, her movements deliberate and practiced. The suit hums softly, circulating air, regulating temperature, monitoring every physiological variable. The silence beyond the hull is absolute. No wind reaches her ears. No sound carries through the thin atmosphere. When her boot finally touches the regolith, the soil compresses slightly beneath her weight, dust rising in slow motion before settling again. The surface has lain undisturbed for nearly four billion years.

"One small step," she whispers instinctively, and then stops herself. Those words belong to another world and another moment. She searches for something new to say, something that acknowledges both the triumph and the burden of this act, but nothing comes. There is only the sound of her breathing and the awareness that humanity has crossed a threshold it can never uncross.

Mars receives its first human footprint without ceremony.

The Mission

The crew's journey begins nine months earlier, timed precisely to the opening of a launch window dictated not by politics or budgets, but by orbital mechanics. Earth and Mars align favorably only once every twenty six months, allowing spacecraft to follow a fuel efficient path known as a Hohmann transfer orbit. The trajectory is not a straight line, but a long, graceful arc around the Sun, trading time for reduced energy demands. Miss the window, and the mission waits more than two years.

Their spacecraft departs Earth atop the most powerful launch system ever assembled, lifting not only the crew vehicle but a sequence of cargo elements that have already been sent ahead. The design favors reuse and volume over elegance, a blunt instrument optimized for hauling mass across interplanetary distances. Inside, the living space feels both expansive and claustrophobic, with high ceilings and wide corridors that cannot disguise the fact that everything is sealed within a metal shell thinner than the walls of a suburban home.

The crew consists of six people selected as much for psychological compatibility as for expertise. Jessica Chen commands the mission, a former astronaut with experience aboard orbital stations and a reputation for calm decision making under pressure. A systems engineer oversees life support, water recycling, and power distribution, carrying responsibility for systems that allow no margin for failure. A geologist serves as the primary scientist, trained to read rocks like archives and landscapes like stories. The medical officer doubles as a botanist, tasked with maintaining both human health and experimental plant systems. A robotics specialist manages autonomous equipment and surface rovers. A pilot rounds out the team, trained to handle landing and ascent under conditions no simulator can fully replicate.

The nine month transit tests them in ways no training fully prepared them for. Radiation exposure accumulates steadily, with galactic cosmic rays and occasional solar events penetrating shielding designed to mitigate, not eliminate, the risk. Dosimeters record each increment, translating invisible particles into numbers that remind the crew of their vulnerability. Psychologically, the distance from Earth becomes tangible. Communication delays stretch from minutes to nearly twenty, turning conversations into exchanges of recorded messages rather than dialogue. Isolation settles in, not as loneliness alone, but as a sense of being untethered from immediacy.

Physically, the absence of gravity takes its toll. Muscles weaken despite rigorous exercise regimens. Bone density decreases as calcium leaches into the bloodstream. The crew floats through the habitat with practiced ease, yet each knows that reentry into gravity will demand painful readjustment. The medical officer monitors biomarkers closely, adjusting nutrition and workload in an ongoing effort to slow deterioration.

The moment Mars begins to grow visibly larger through the observation window marks a psychological shift. What had been an abstract destination becomes a presence. The planet's colors sharpen. Its surface details emerge. The crew gathers often at the window, drawn by a mixture of anticipation and unease. Mars does not look welcoming. It looks exposed.

The landing sequence compresses months of anticipation into seven minutes of controlled chaos. As the spacecraft enters the atmosphere, friction heats the exterior to temperatures that would melt steel. Automated systems adjust orientation, deploying aeroshells and decelerators with precise timing. The thin Martian atmosphere provides little margin for error, forcing a choreography of parachutes, retrorockets, and final powered descent. Every second is scripted, yet every second carries risk.

The target is Jezero Crater, chosen for its scientific promise and relative safety. Once a lake fed by a river delta, the crater contains layered sediments formed in the presence of liquid water. Clay minerals line its basin, preserving chemical signatures that on Earth often correlate with life. Radar surveys indicate accessible subsurface ice, a critical resource for any sustained human presence. The terrain is relatively flat, minimizing landing hazards while maximizing scientific return.

When the spacecraft settles onto the surface, the crew waits in silence as systems stabilize. Pressure equalizes. Sensors confirm structural integrity. Only then does Jessica Chen give the order to begin surface operations. The first extravehicular activity is procedural rather than celebratory. Suits are checked. Airlocks cycled. Tools secured. The habitat module, delivered by an earlier cargo mission, stands nearby like a waiting sentinel.

Stepping onto Mars is both exhilarating and exhausting. Movement feels awkward under reduced gravity, each step carrying the potential for imbalance. The suits restrict motion, turning simple actions into deliberate sequences. Dust coats everything, clinging electrostatically to surfaces and joints. The crew establishes routines quickly, knowing that familiarity breeds efficiency and efficiency breeds safety.

Infrastructure becomes their immediate focus. Solar arrays unfold, angling themselves to capture the weak sunlight. A compact nuclear power unit provides baseline energy through dust storms and night cycles. Atmospheric processors draw in carbon dioxide rich air, separating oxygen for breathing and fuel production. Water extraction units begin heating regolith to release ice trapped below the surface. Greenhouse modules are activated, their first crops symbolic as much as practical.

Communication with Earth settles into its new rhythm. Messages are sent and received on delay, each carrying the weight of isolation. The crew maintains detailed logs, documenting not only technical performance but emotional states and interpersonal dynamics. These records will become as valuable as any rock sample.

Waste management emerges immediately as a critical concern. There is no away on Mars. Every byproduct must be stored, processed, or repurposed. On Earth, waste disappears into systems built over centuries. Here, it accumulates visibly, a reminder that sustainability is not optional.

The mission unfolds in sols rather than days, each Martian rotation slightly longer than Earth's. Sol One brings exhilaration tempered by exhaustion. Sol Two introduces the first minor malfunctions. Sol Three establishes routine. With each passing sol, Mars becomes less alien and more intimate, revealing both its hostility and its fragile beauty.

The footprints left behind are light, but their meaning is heavy. Humanity is no longer observing Mars from afar. It is inhabiting it, however briefly, and every action carries consequence.

The First Environmental Decisions

The first ethical reckoning arrives not with a dramatic discovery, but with a mundane necessity. Waste accumulates faster than anyone anticipated. Human bodies produce it constantly. Packaging piles up as equipment is unpacked. Filters clog. Cleaning agents leave residue. On Earth, waste is a background problem managed by invisible infrastructure. On Mars, it is a visible and unavoidable presence inside a habitat smaller than a modest apartment.

The crew gathers around the central table on Sol Six to discuss disposal. The systems engineer outlines the inventory with professional detachment, listing biological waste, graywater, damaged components, and chemical byproducts. Every category presents risk. Burying waste removes it from sight, but not from the planet. Incineration reduces volume, but releases compounds into an atmosphere that does not readily disperse them. Storing waste for return to Earth preserves Mars, but consumes mass and energy the mission can barely spare. Recycling offers the cleanest option, but pushes life support systems beyond their tested limits.

Mission Control's delayed messages emphasize practicality. Mars, they argue, is sterile. There is no ecosystem to poison. The priority is crew survival and mission success. The geologist counters that sterility is an assumption, not a fact. Subsurface environments remain unexplored. Contamination introduced now would persist indefinitely. The medical officer notes that water recovery efficiency must approach near perfection regardless of disposal choices, making recycling unavoidable in practice.

Jessica Chen listens without interruption, weighing not only the arguments but their implications. Apollo missions left waste on the Moon without consequence because the Moon offered no plausible habitat for life. Mars is different. Its history includes rivers, lakes, and possibly biology. The commander decides on a hybrid approach. Biological waste will be fully processed and recycled. Chemical waste will be stored and cataloged for return when possible. No burial will occur without explicit justification and documentation. The decision is imperfect, but it establishes intent. Mars will not be treated as a dumping ground by default.

Water forces a second confrontation. On Sol Twelve, radar mapping confirms a substantial ice deposit beneath the regolith roughly ten meters from the habitat. The volume is staggering by mission standards, enough to support years of habitation. The discovery triggers immediate excitement and unease. Water means survival, expansion, and independence from Earth. It also represents a pristine archive of Martian climate history and a potential refuge for extant life.

The crew debates extraction methods late into the sol. Conservative access would minimize disturbance but limit supply. Industrial extraction would meet future needs but irreversibly alter the site. Shipping water from Earth is technically possible but economically absurd. The geologist emphasizes that ice cores could reveal millions of years of climate data. The astrobiology consultants on Earth warn that subsurface ice could harbor microbial life protected from radiation. The engineer reminds everyone that dehydration kills quickly.

The compromise mirrors the earlier decision. Minimal extraction begins with extensive monitoring. Samples are taken methodically. Extraction rates are capped. The ice is treated as a shared heritage rather than a commodity. The choice keeps the crew alive without committing Mars to immediate exploitation. It also sets a precedent that may not survive future missions.

Soil sampling presents a quieter but more insidious risk. Strict sterilization protocols slow exploration dramatically. Relaxing them accelerates science while increasing contamination. The Viking landers of 1976 were sterilized so thoroughly that they may have destroyed the organic molecules they sought to detect. The crew is keenly aware of this paradox. Over protection can blind discovery. Under protection can erase it.

The crew adopts tiered protocols. Regions deemed high priority for bio signature preservation are approached with maximal caution. Less sensitive areas allow relaxed procedures. Every deviation is logged. Every sample is contextualized. The work is painstaking, and progress feels slow, but the alternative feels reckless.

Political pressure surfaces unexpectedly on Sol Nineteen. Messages arrive urging symbolic acts of ownership. Flags are suggested. Markers proposed. Corporate sponsors request recognition for their investment. International partners remind the crew of treaty obligations that prohibit national appropriation. The debate is not abstract. Symbols shape narratives. Narratives shape policy.

The crew chooses restraint. A mission emblem is placed near the habitat, representing humanity rather than any nation. No territorial claim is made. The act is subtle, but its implications ripple outward. Mars will not be claimed with the first footprint, at least not by this crew.

None of these decisions feel final. They feel provisional, fragile, and vulnerable to reversal. Yet together they mark the emergence of an ethic, tentative and contested, but real.

What We Found

Mars reveals itself slowly, resisting easy interpretation. The landscape around Jezero Crater tells a story written in layers. Ancient river channels cut through rock, their curves unmistakable even after billions of years. Sedimentary strata stack like pages, each representing a chapter of environmental change. Dust devils spin across plains, tracing invisible currents in an atmosphere too thin to carry sound.

Olympus Mons looms on the horizon during clear sols, its sheer scale defying intuition. The largest volcano in the solar system rises so high that its summit pierces the thin atmosphere, yet its slopes are gentle, shaped by lava flows rather than explosive eruptions. Phobos and Deimos arc overhead, irregular and intimate, closer than any moon seen from Earth.

Sunsets disorient the crew. The sky shifts to pale blue near the horizon, the opposite of Earth's red sunsets, caused by dust scattering light differently in the thin air. The effect is beautiful and unsettling, a reminder that familiarity cannot be assumed even in something as universal as twilight.

The first samples excite and frustrate in equal measure. Hematite concretions appear in abundance, small spheres formed in the presence of water. Clay minerals such as montmorillonite confirm prolonged contact with liquid water. Carbonates suggest neutral pH environments conducive to life. Perchlorates complicate analysis, toxic to humans yet indicative of active chemistry.

Organic molecules appear in trace amounts, complex carbon based compounds that could originate from biological processes or arrive via meteorites. Methane plumes detected intermittently in the atmosphere deepen the mystery. Seasonal variations suggest dynamic processes, but whether they are biological or geological remains unresolved.

The scientific debate intensifies. Astrobiologists caution against premature conclusions. Geo chemists offer abiotic explanations. Carl Sagan's standard looms large, reminding everyone that extraordinary claims require extraordinary evidence. Yet absence of evidence is not evidence of absence. Mars refuses to be categorized neatly as dead or alive.

As data accumulates, a broader realization takes hold. Mars once resembled Earth in fundamental ways. It possessed liquid water, a thicker atmosphere, and a magnetic field that shielded it from solar radiation. Then something changed. The magnetic field weakened. The solar wind stripped the atmosphere. Water froze or escaped. Surface conditions became hostile. The planet did not explode or shatter. It faded.

The possibility that life retreated underground lingers. Subsurface environments could still harbor liquid water heated by geothermal energy and protected from radiation. If life persists, it does so quietly, adapted to scarcity and isolation.

The emotional impact on the crew is profound. Mars feels less like a blank canvas and more like a patient whose history is written into its scars. Disturbing it now feels intrusive. Studying it feels necessary. The tension between curiosity and care sharpens daily.

The Choice Point

On Sol Forty Seven, the crew convenes for a meeting that everyone understands will shape more than the remainder of the mission. New directives from Earth arrive after weeks of anticipation. Corporate sponsors push for accelerated resource extraction to demonstrate viability. Scientific institutions urge caution, citing unresolved questions about life and preservation. Political actors debate recognition and control.

The cramped habitat amplifies the intensity of the discussion. Jessica Chen opens the floor, framing the issue plainly. Humanity is here. More missions will follow. What precedent will be set.

The argument for aggressive expansion emphasizes survival and momentum. Mars is framed as lifeless. Resources are abundant. Delay risks losing public support and investment. Terra forming is invoked as a long term goal, requiring early commitment.

The conservative position counters with humility. The planet is barely understood. Contamination risks are irreversible. Science requires patience. Preservation must come before exploitation.

A middle path emerges reluctantly. Sustainable exploration, minimal impact, strict oversight, and international cooperation. Take what is necessary to survive, but no more. Treat Mars as a shared responsibility rather than a possession.

Personal histories surface. Jessica Chen thinks of her children and the future they will inherit. The geologist grapples with the shift from studying Mars as an object to encountering it as a place. The engineer recalls environmental disasters on Earth that began with good intentions and ended in regret.

The decision does not resolve the conflict. It defers it. Option C is chosen for now, with full awareness that six people cannot decide the fate of a planet. What they can do is demonstrate an alternative to haste.

Jessica records her log late that sol, her voice steady but reflective. They have chosen caution over speed and preservation over profit, knowing that their influence is limited. The future of Mars will be shaped by many who follow. The hope is that someone is paying attention.

Watching Worlds Watch Us

From classrooms to laboratories, from corporate boardrooms to assembly halls, millions watch humanity's first sustained steps on Mars. Children trace the rover paths on screens. Scientists parse data streams. Investors calculate returns. Diplomats debate language.

Mars becomes a mirror held up to Earth. The question is not whether humans can reach other worlds, but whether they can arrive without repeating familiar patterns. Exploration and exploitation have long been intertwined. Untangling them now is both necessary and unprecedented.

The footprints on Mars are light, but their implications are heavy. They mark not conquest, but encounter. Whether that encounter becomes a relationship or a repetition remains unresolved.

As the world celebrates, attention shifts elsewhere. While Mars captures imagination, quieter transformations unfold closer to home. The Moon, humanity's nearest neighbor, is already being altered, mined, and claimed with little scrutiny.

While we argued about Mars, we began reshaping the Moon.

And almost no one noticed.

5

The Moon We're Already Ruining

While we dreamed of Mars, we were already destroying the Moon. Unlike Mars, two hundred and forty million kilometers away, the Moon is close enough to see with the naked eye. It hangs in the night sky as it has for all of human history, unchanged in outline, familiar in its cycles. Children point to it instinctively. Poets still write about it. Sailors once navigated by it. What is new is not the Moon itself, but what humans are now doing to it.

The question that will confront future generations is simple and uncomfortable. When our children look up and say, "We did that," will they say it with pride or with shame.

The Current Lunar Rush

The modern rush to the Moon is not a prediction. It is a matter of record. After decades of relative quiet following the end of the Apollo era, lunar exploration accelerated rapidly in the late 2010s and early 2020s. What had once been the domain of two superpowers became a crowded field of nations and private companies, each pursuing its own priorities with minimal coordination.

In 2019, China's Chang'e 4 mission achieved the first landing on the far side of the Moon, a technical milestone that demonstrated sophisticated navigation and communications capabilities. Four years later, in 2023, India's Chandrayaan 3 successfully landed near the lunar south pole, placing instruments within reach of regions long suspected to contain water ice. In 2024, multiple private landers attempted missions under commercial contracts, some succeeding and others failing, each adding hardware, debris, and disturbance to the lunar surface.

By the mid 2020s, the pace accelerated further. NASA's Artemis program aimed to return humans to the Moon by 2025 or 2026, marking the first crewed lunar landing since 1972. China announced plans for a permanent lunar research station by 2030, in partnership with Russia. India prepared additional sample return missions. Japan supported commercial lunar ventures. South Korea and the United Arab Emirates joined the growing list of nations with active lunar ambitions. Private companies outlined plans for cargo delivery, infrastructure, tourism, and resource extraction.

The difference between this era and Apollo was not simply technological. It was philosophical. Apollo missions were brief visits driven by geopolitical competition and symbolic achievement. Flags were planted, experiments deployed, and crews departed. Today's missions were framed explicitly around permanence. The language shifted from exploration to utilization. The Moon was no longer a destination. It was a foothold.

This shift was most evident in how missions clustered geographically. Nearly every major space faring actor targeted the same region: the lunar south pole. Unlike the equatorial sites favored during Apollo, the south pole offered a rare and valuable combination of conditions. Permanently shadowed craters near the pole never received direct sunlight, preserving water ice delivered by comets or formed through interactions with the solar wind. Adjacent ridges received near constant sunlight, ideal for solar power. Together, these features created the most valuable real estate beyond Earth.

Shackleton Crater became the centerpiece of this convergence. Its rim offered elevated terrain bathed in sunlight, while its interior remained in permanent darkness, cold enough to trap volatiles for billions of years. Nearby craters such as Haworth, Cabeus, and Shoemaker exhibited similar properties. The total area of these permanently shadowed regions was small by planetary standards, roughly comparable to the footprint of a modest city on Earth. Yet more than a dozen nations and companies expressed interest in accessing them.

There was no global mechanism to allocate access, assess cumulative impact, or enforce restraint. Each actor planned independently, guided by national law, corporate strategy, or political ambition. The result resembled a gold rush town before the arrival of governance. Claims were implied through presence. Infrastructure was prepositioned. Landing sites were informally marked as future zones of operation.

Activity on the surface intensified quickly. Rovers traversed regolith that had formed over millions of years through micrometeorite bombardment. Landers fired engines that blasted dust and rock across kilometers. Instruments were deployed, cables laid, experiments activated. Each action was individually modest. Collectively, they altered the Moon's surface in ways that were irreversible on any human timescale.

Some changes were unintentional. In 2019, the privately funded Beresheet lander crashed during its descent, scattering debris and, controversially, microscopic tardigrades onto the lunar surface. Though the organisms were dehydrated and likely inactive, their presence underscored the absence of enforceable planetary protection standards for the Moon. In 2023, another commercial lander failed, adding a new debris field to an already cluttered environment. Other impacts were deliberate, conducted to test landing systems or seismic responses, each leaving scars that would persist indefinitely.

The Moon lacks an atmosphere capable of erasing tracks or redistributing material. There is no wind to smooth disturbances, no rain to settle dust, no tectonics to recycle the surface. Footprints, rover tracks, and blast zones endure for millions of years. Apollo boot prints remain as crisp today as they were in 1969. Every new mission adds another permanent mark.

Despite this, there is no requirement for environmental impact assessments before lunar operations. There are no exclusion zones beyond voluntary guidelines. There is no traffic control system to manage landings or protect sensitive sites. The Outer Space Treaty of 1967 establishes lofty principles about peaceful use and shared heritage, but it offers little practical guidance for managing cumulative harm.

Dr. Aisha Patel watches this unfold with growing unease. A lunar geologist by training, she built her career studying Apollo samples, extracting insights from a handful of rocks collected half a century earlier. Those samples transformed understanding of the Moon's formation and early solar system history. For decades, scientists waited for a return that would expand this record carefully and deliberately.

What they received instead was acceleration without pause. Aisha speaks often at conferences, urging restraint and coordination. She asks a simple question that rarely finds traction. We waited fifty years to return. Can we wait ten more to do it right.

Across the table from her in policy discussions sits Marcus Chen, a mining executive whose optimism is as genuine as Aisha's concern. He sees the Moon not as a fragile archive, but as a solution. Lunar resources, he argues, could unlock clean energy and reduce pressure on Earth's ecosystems. Delay, in his view, is itself unethical. We are not destroying the Moon, he insists. We are saving Earth with its resources.

Both perspectives carry weight. Both are grounded in legitimate concerns. What is missing is a framework capable of holding them together without defaulting to the loudest or fastest actor.

As missions multiply and infrastructure spreads, the window for thoughtful stewardship narrows. The Moon is being shaped not by a single decision, but by accumulation. Each landing, each track, each experiment contributes to a trajectory that is becoming harder to reverse.

The rush is not coming. It is already here.

What We're Doing Wrong

The most consequential failure unfolding on the Moon is not technological. It is institutional. Humanity has entered a phase of permanent lunar activity without building the governance structures required to manage it responsibly. The result is a vacuum in which ambition outruns coordination and precedent is set by whoever arrives first.

The Outer Space Treaty of 1967 remains the foundational document governing activity beyond Earth. Drafted during the Cold War, it establishes that space shall be used for peaceful purposes and that celestial bodies are the province of all humankind. It prohibits national appropriation by claim of sovereignty, use, or occupation. These principles are noble, but they are also vague. The treaty does not define peaceful purposes. It does not address environmental protection. It does not establish enforcement mechanisms. It was written for an era of flags and footprints, not permanent bases and industrial extraction.

In practice, the absence of clarity has produced friction. Missions operate independently, sometimes interfering with one another's scientific objectives. Instruments sensitive to vibration or dust have been affected by nearby landings. Radio telescopes attempting to study the quiet lunar far side face interference from communication systems. Nations and companies stake informal claims by landing near valuable resources, creating de facto control without legal ownership.

The United Nations Committee on the Peaceful Uses of Outer Space exists to facilitate coordination, but its authority is limited. It issues recommendations rather than requirements. Decisions require consensus among dozens of nations with divergent interests. Progress is slow by design, while lunar activity accelerates rapidly. The mismatch between pace and process leaves the Moon exposed.

Rocket exhaust introduces another layer of harm. Although the Moon lacks a traditional atmosphere, it possesses an exosphere, an extremely thin envelope of particles constantly replenished and lost through interactions with solar wind, micrometeorite impacts, and radioactive decay. This exosphere is a scientific treasure. It offers insight into how airless bodies interact with space, preserving processes that have remained largely unchanged for billions of years.

Rocket landings disrupt this delicate system directly. Exhaust plumes strike the surface at high velocity, ejecting regolith and injecting exhaust gases into the exosphere. Apollo missions produced detectable changes that persisted for months. Modern landers are larger and more numerous. Modeling studies suggest that cumulative effects could alter the composition and behavior of the exosphere in measurable ways, particularly near frequently used landing sites.

These changes matter because once altered, the exosphere cannot be restored. There is no mechanism to reset it to its original state. Future scientists lose the opportunity to study pristine conditions. The laboratory is altered by the act of observation.

Regolith disruption compounds the damage. Lunar regolith is not simply dirt. It is a layered archive formed through billions of years of micrometeorite bombardment and space weathering. Each layer records a different era of solar system history. Solar wind particles implanted within grains preserve information about past solar activity. Tiny glass beads formed by impacts encode thermal histories.

Rover tracks, landing blast zones, and excavation permanently disturb these layers. Unlike Earth, the Moon has no weather to erase scars. Tracks remain until buried by future impacts, a process that takes millions of years. Construction for bases and mining would require moving vast quantities of material, destroying context and mixing layers irreversibly.

The permanence of damage extends to cultural heritage. The Moon already hosts the most significant archaeological sites beyond Earth. Apollo landing sites represent humanity's first steps onto another world. Soviet landers and rovers tell parallel stories of exploration. Chinese missions add new chapters. Together, they form a record of human expansion into space.

Yet these sites enjoy no meaningful protection. There are no exclusion zones enforced internationally. Voluntary guidelines exist, but they bind only those who choose to follow them. Domestic laws, such as those enacted by the United States, apply only to national entities. A spacecraft from another nation or company could land close enough to damage artifacts without violating any binding international rule.

The Apollo 11 site, perhaps the most iconic location in space history, is particularly vulnerable. As traffic increases near the lunar south pole and equatorial regions, the risk of accidental damage grows. Tourism proposals raise further concerns. A single careless landing could destroy artifacts that cannot be replaced.

This loss would not be merely symbolic. It would sever a tangible link to humanity's early space faring history, equivalent to bulldozing ancient ruins on Earth. The absence of protection reflects not malice, but neglect. It is easier to act than to pause.

The Helium 3 Gold Rush

Underlying much of the lunar rush is a powerful economic narrative centered on helium three. Helium three is a rare isotope on Earth, consisting of two protons and one neutron. It is produced in small quantities through the decay of tritium and is not naturally abundant. On the Moon, however, billions of years of exposure to the solar wind have implanted helium three into the regolith, particularly in the upper few meters.

The appeal of helium three lies in its theoretical potential as a fusion fuel. Fusion reactions involving helium three and deuterium produce large amounts of energy with fewer radioactive byproducts than more commonly studied fusion pathways. In principle, helium three fusion could provide clean, abundant energy. In practice, it remains unproven. Despite decades of research, controlled fusion has yet to produce net energy at commercial scale. Helium three fusion is even more technically challenging than existing approaches.

Nevertheless, the promise is seductive. Proponents argue that a single cargo of helium three could power a nation for a year. Estimates of its market value reach into the trillions of dollars. For nations concerned about energy security and climate change, the allure is obvious. Lunar helium three becomes a symbol of salvation.

This promise drives policy even in the absence of proof. China has openly cited helium three as a strategic interest in lunar exploration. Russia partners in pursuit of shared infrastructure. Private companies position themselves as pioneers of extraterrestrial resource extraction. Legal ambiguity fuels momentum. The Outer Space Treaty prohibits national appropriation, but it does not explicitly forbid resource ownership. Domestic laws in the United States and Luxembourg assert that companies may own what they extract, creating a patchwork of claims without global consensus.

The environmental cost of helium three mining is staggering. Concentrations in regolith are measured in parts per billion. To extract a single kilogram requires processing hundreds of thousands of tons of material. Achieving quantities sufficient to power Earth would require strip mining vast areas of the lunar surface, heating regolith to high temperatures, and disposing of depleted waste.

Such operations would scar the Moon on a scale visible from Earth with telescopes. They would destroy the scientific record embedded in regolith layers. They would alter surface reflectivity and thermal properties. The accumulation of waste material would pose new hazards.

Perhaps most troubling is that no comprehensive risk assessment exists. There is no requirement to demonstrate that helium three fusion is achievable before committing to extraction. There is no global process to evaluate whether the benefits outweigh the costs. Decisions are driven by hope and competition rather than evidence and caution.

Marcus Chen acknowledges these challenges but remains convinced. He argues that humanity cannot afford to leave potential solutions unexplored. Climate change demands bold action. If the Moon can help save Earth, he believes, the trade off is justified. His perspective resonates with many who fear that delay equals doom.

Dr. Aisha Patel counters that this framing presents a false choice. Earth has viable renewable energy pathways that do not require destroying another world. Helium three mining risks repeating the same pattern of externalizing costs, this time onto a celestial body that cannot recover. Saving Earth by sacrificing the Moon, she argues, reveals a failure to imagine solutions that respect limits.

Between these positions lies a silence where governance should be.

What We're Losing

The Moon is more than a storehouse of resources. It is a scientific archive unmatched in the solar system. Without plate tectonics or weather, its surface preserves a record of early solar system history that Earth erased long ago. Impact craters record bombardment rates. Regolith layers store solar wind signatures. Meteorites from Earth itself may lie preserved, offering clues to early terrestrial conditions.

Disturbing this archive erases information irretrievably. Science conducted without foresight precludes future questions not yet imagined. Control sites, areas left untouched for comparison, are essential for understanding change. Without them, interpretation becomes guesswork.

Beyond science lies culture. The Moon occupies a unique place in human imagination. It has guided calendars, inspired myths, and shaped art for millennia. It has been a symbol of constancy and mystery. Industrializing it risks diminishing that shared meaning. Visible scars would alter how future generations perceive the night sky, transforming a symbol into a reminder of unchecked expansion.

Unlike Earth, the Moon cannot heal. There are no ecosystems to regenerate, no processes to erase damage. What is lost is lost permanently. This irreversibility demands restraint.

The analogy is unavoidable. Mining the Moon without protection is akin to drilling inside a cathedral or carving into ancient monuments. Utility does not negate value. Some places matter because they exist, not because they can be exploited.

The question becomes whether humanity can recognize intrinsic value beyond immediate use. Whether some places can be preserved simply because they connect us to history, to science, to wonder.

A Mirror in the Night Sky

A child looks up at the Moon and asks what it is made of. The answer today includes dust, rock, and history. Tomorrow, it may include scars.

This transformation is not distant. It is unfolding now. Missions planned for the next decade will determine whether the Moon becomes a managed commons or an unregulated frontier. The window for protection is closing quickly.

Twelve humans walked on the Moon between 1969 and 1972. By the mid twenty first century, hundreds or thousands may follow. The difference is not numbers, but intent. Apollo was exploration. What comes next risks becoming exploitation.

The Moon is the testing ground. How humanity treats it will reveal whether it has learned anything from Earth. Beyond the Moon lie worlds even more fragile and potentially alive. There, the stakes are higher still.

Because there, we might not be alone.

6

When We Found Water (And What We Did With It)

The image appeared on screens worldwide at 11:47 UTC. In control rooms and laboratories, in classrooms and living rooms, people leaned closer without realizing they had moved. On the screen was Europa, Jupiter's ice moon, illuminated faintly by reflected light from the giant planet it orbited. Against that dim glow rose something extraordinary. A plume of water vapor erupted from the moon's fractured surface, arcing into space like a frozen geyser caught in mid breath. The plume climbed nearly two hundred kilometers above the ice, towering twenty times higher than Mount Everest, its structure resolved in detail no one had dared hope for.

Within seconds, spectroscopic data overlaid the image, translating light into chemistry. Scientists watched the lines appear, one by one, as software parsed the signal. Water, unmistakable. Salts. Then carbon based compounds, not just simple methane, but more complex organics. Molecules that on Earth served as the scaffolding of biology.

For a long moment, no one spoke. Europa Clipper was on its forty third flyby of the moon, executing a mission planned decades earlier with the explicit goal of understanding habitability. It was not designed to find life. Yet here it was, revealing an ocean sealed beneath ice, twice the volume of all Earth's oceans combined, warmed by tidal forces and chemically active.

An ocean that might not be empty.

Humanity had searched for water beyond Earth for generations, driven by the belief that where there was water, there might be life. Now that search had crossed a threshold. The universe had not merely offered a hint. It had offered a flood.

The discovery forced a reckoning that extended far beyond Europa itself. It challenged assumptions about uniqueness, scarcity, and entitlement. It asked not only whether life existed elsewhere, but what humanity would do if it did.

The Ocean Worlds

For most of modern science, Earth was considered an anomaly. It was the only known world with vast oceans, a stable climate, and life that had persisted for billions of years. Early observations of the solar system reinforced this belief. The inner planets were barren or hostile. The outer planets were gaseous giants without solid surfaces. Water seemed rare and fragile.

That picture unraveled slowly. Europa was first observed in 1610 by Galileo Galilei, one of four large moons whose regular motion helped confirm the heliocentric model. For centuries, it was little more than a point of light. It was not until the Voyager missions flew past Jupiter in 1979 that Europa's surface revealed its true nature. Instead of being heavily cratered like the Moon, it appeared smooth and fractured, crisscrossed by dark lines that suggested movement beneath the ice.

The Galileo spacecraft, orbiting Jupiter between 1995 and 2003, transformed suspicion into evidence. As it passed Europa repeatedly, instruments detected subtle variations in Jupiter's magnetic field. The only plausible explanation was a conductive layer beneath Europa's surface. Salty liquid water. A global ocean.

Subsequent observations refined the picture. Europa's ice shell was estimated to be between ten and thirty kilometers thick. Beneath it lay an ocean between sixty and one hundred fifty kilometers deep, resting on a rocky mantle. The amount of water was staggering, roughly twice the volume of all Earth's oceans combined. Tidal heating from Jupiter's gravity prevented the ocean from freezing solid, while surface features suggested exchange between the ocean and the ice shell.

Europa was not alone.

In 2005, the Cassini spacecraft observed geysers erupting from the south pole of Enceladus, a small moon of Saturn once dismissed as inert. Jets of water vapor and ice particles shot hundreds of kilometers into space. Cassini flew through these plumes, sampling them directly, and found salts, organic molecules, and molecular hydrogen, strong evidence of hydrothermal activity on the ocean floor. Enceladus possessed a global ocean beneath its ice, maintained by tidal flexing despite its small size.

Titan, Saturn's largest moon, expanded the definition of habitability further. Larger than Mercury, Titan had a thick atmosphere rich in nitrogen and methane. Lakes and seas of liquid hydrocarbons pooled on its surface. Beneath that alien landscape, models indicated a subsurface water ocean. Titan's chemistry was extraordinarily rich, producing complex organic molecules that rained onto the surface, creating a laboratory for prebiotic chemistry unlike anything on Earth.

The list continued to grow. Ganymede, the largest moon in the solar system, showed evidence of multiple stacked oceans beneath its surface. Callisto appeared to harbor a buried sea. Mimas, long thought to be a dead world, revealed signs of an ocean beneath its cratered shell. Even Pluto, distant and cold, showed evidence of a subsurface ocean insulated by ice.

The realization was profound. Earth was not unique in possessing oceans. It was merely unique in displaying them openly. Across the outer solar system, hidden oceans vastly expanded the habitable volume available for life. There was more liquid water beneath ice moons than in all of Earth's seas combined.

Chemistry reinforced the case. These oceans contained water, salts, and organic molecules. Energy sources were abundant. Tidal heating generated warmth. Chemical reactions between water and rock released energy. Hydrothermal vents, analogous to those discovered on Earth's ocean floor in 1977, were not just possible but likely.

On Earth, hydrothermal vents supported entire ecosystems in complete darkness. Giant tube worms, blind shrimp, and specialized bacteria thrived without sunlight, powered by chemical energy rather than photosynthesis. Life proved adaptable beyond imagination.

Europa's ocean offered the same ingredients. Liquid water. Organic chemistry. Energy. Time measured in billions of years. The checklist for life appeared complete.

The universe, it seemed, was wet.

The Life Question

The possibility of life beneath Europa's ice reshaped scientific imagination, but it also forced humility. Life on Earth proved astonishingly adaptable. Organisms thrived in boiling springs, frozen Antarctic ice, hyper saline lakes, acidic pools, and radioactive waste. Bacteria were found kilometers beneath the surface, metabolizing rock itself. Extremes were not barriers. They were niches.

These discoveries re framed habitability. Pressure, darkness, and cold ceased to be disqualifies. Europa's ocean, sealed away from sunlight and subjected to immense pressure, might be benign compared to environments life already occupied on Earth.

Most scientists expected that any life discovered beyond Earth would be microbial. Single celled organisms required fewer evolutionary steps and adapted readily. Yet speculation persisted about complexity. If energy gradients were stable and abundant, could multicellular life arise. Could ecosystems form. Could something move through Europa's dark ocean.

Even microbial life would be trans formative. It would answer one of humanity's oldest questions. Are we alone. It would demonstrate that life arose readily when conditions allowed. Biology would become a universal science rather than an Earth bound one.

More unsettling was the possibility of a second genesis. If life on Europa arose independently, using different genetic structures or alternative chemistry, it would imply that life was not a rare accident but a common outcome of planetary evolution. The universe would not merely permit life. It would produce it.

This possibility carried ethical consequences as significant as the scientific ones. Discovering life elsewhere would impose responsibility. The moment organic molecules appeared in Europa's plume data, planetary protection protocols moved from abstract to urgent.

Every spacecraft carried Earth microbes. Perfect sterilization was impossible. Some bacteria survived vacuum, radiation, and extreme temperatures. Tardigrades, microscopic animals, could endure conditions that would kill most life. Surveys consistently found microbes even in the cleanest assembly facilities.

Forward contamination posed a genuine threat. If a probe penetrated Europa's ice and carried Earth microbes into the ocean, the consequences would be irreversible. Those microbes might die. Or they might survive and out compete native life, altering or destroying an alien biosphere before it was even discovered. Scientific value would be lost forever. If Europa harbored life, contamination could amount to extinction on a planetary scale.

Backward contamination raised different fears. If samples were returned to Earth, could alien organisms pose a threat. Most scientists considered catastrophic scenarios unlikely, but uncertainty remained. Earth life had never encountered alien biology. Caution was prudent.

Existing planetary protection protocols struggled to keep pace. Guidelines existed, but enforcement was weak. Commercial missions often fell outside regulatory frameworks. No binding international law governed the treatment of potentially living worlds.

The ethical asymmetry was stark. Failing to detect life was a loss. Destroying life before discovering it was a catastrophe. This imbalance argued for restraint. Yet restraint faced political and economic pressure. Delay risked losing funding. Competition rewarded speed.

Europa became a mirror, reflecting not only the possibility of alien life but humanity's unresolved values.

The Ethical Bomb

The discovery of water and organics beneath Europa's ice detonated an ethical bomb that could not be defused by technical solutions alone. It posed a dilemma akin to the trolley problem, scaled to cosmic proportions.

Imagine that microbial life is confirmed in Europa's ocean. Imagine that the same ocean contains vast quantities of water that could be split into hydrogen and oxygen to fuel spacecraft. Extracting that water could enable deep space missions, asteroid deflection, and perhaps even planetary defense. It could save human lives. It would also destroy habitat and possibly extinguish an entire alien biosphere.

Do we exploit the resource. Do we protect the life. Can we do both.

Anthropocentric ethics argued that human needs took priority. Humans were the only beings capable of moral reasoning that we knew of. Microbes did not suffer. Earth's bacteria were killed routinely for human benefit. Why should alien microbes be different.

Biocentric perspectives countered that all life had intrinsic value. Life might be rare enough in the universe to be sacred. Destroying an entire biosphere for convenience was indefensible.

Ecocentric frameworks focused on systems rather than individuals. The question was not whether microbes mattered, but whether the ecosystem could be studied and preserved while meeting limited human needs. Stewardship replaced ownership.

Utilitarian approaches attempted calculation, weighing benefits and harms. Yet assigning value to alien life proved impossible. Whose utility counted.

Present humans. Future humans. Unknown organisms.

Deep ecology rejected the premise entirely, arguing that nature had rights independent of human use.

The hardest aspect was emotional. Microbes were invisible. They did not inspire empathy. Yet their significance was unprecedented. They represented proof that life was not unique to Earth.

Who had the authority to decide their fate. Governments. Corporations. Scientists. Humanity as a whole. Future generations. No framework existed.

The stakes were the highest humanity had ever faced.

The Corporate Race

Water was not just a scientific discovery. It was an economic revolution waiting to happen. Water could be split into hydrogen and oxygen, the most efficient rocket propellant known. Launching fuel from Earth was expensive. Mining water in space promised a self sustaining space economy.

Europa's water reserves were vast. In theory, it could become a refueling station for deep space missions. The outer solar system's most valuable real estate.

Space agencies framed missions as science, but resource potential was noted quietly. Private companies, though unable to reach Europa yet, positioned themselves for a future where extraction was possible. The pattern was familiar. Exploration. Assessment. Utilization. Exploitation.

Corporate logic was relentless. First mover advantage mattered. Competition rewarded speed. Environmental protection slowed progress. Governments struggled to regulate industries they relied on for innovation and funding.

The debate played out in laboratories and boardrooms. An astrobiologist warned that drilling into Europa's ice was like burning the Library of Alexandria for warmth. An engineer replied that water could fuel missions that might save humanity from extinction.

Neither was a villain. Both were right in part.

The Line We Have Not Crossed Yet

Europa's ocean remains sealed beneath ice, untouched by drills. For now. The next two decades will determine whether that restraint holds.

Two futures diverge. In one, Europa is penetrated, sampled aggressively, and exploited. No life is found, or it is destroyed before discovery. The pattern continues outward. In the other, Europa is studied carefully, life is protected, and humanity demonstrates that it has learned from Earth.

The choice is coming quickly.

Water on distant moons. Ice on Mars. Minerals on asteroids. The resources are there. The race is accelerating.

The question is not whether we will reach them. It is whether we will do so as conquerors or caretakers.

And who decides which we become.

7

Why This Would Be Profound

If humanity were to discover even the simplest microbial life beneath Europa's ice, the consequences would be immediate and irreversible, not because of what those organisms could do, but because of what they would mean. One of humanity's oldest questions would no longer be philosophical. It would be empirical. The question of whether we are alone would be answered not with radio signals from distant stars, but with chemistry drawn from an alien ocean within our own solar system. Life would no longer be a singular miracle confined to Earth. It would be a natural outcome of planetary processes given the right conditions.

That realization would force a shift in worldview as profound as the Copernican revolution. Humanity would move away from an anthropocentric understanding of the cosmos, where life was centered on Earth and exceptional by default, toward a biocentric one, where life was woven into the fabric of the universe. Earth would no longer be the living planet, but one of many. Life would cease to be a cosmic accident and instead become something closer to a cosmic imperative. The scientific conversation would shift decisively from whether life exists elsewhere to how common it is and what forms it takes.

The scientific implications would be equally transformative. Biology would no longer be a discipline constrained to a single data point. Comparative biochemistry would become possible, allowing scientists to study life that did not share Earth's evolutionary history. Assumptions about DNA, amino acids, and metabolism would be tested rather than taken for granted. Evolution itself could be examined across independent origins, revealing which features of life were contingent and which were inevitable. The discovery would not merely expand biology. It would redefine it.

Yet the most destabilizing consequences would be ethical rather than scientific. If life were common, then humanity's relationship to the universe would change fundamentally. The moral status of life would no longer be limited to what we could see, hear, or empathize with. The question would no longer be whether alien microbes mattered, but whether humanity had the right to treat them as disposable simply because they were small, silent, and unfamiliar.

This is where wonder gives way to unease.

The Contamination Nightmare

The danger facing Europa is not theoretical. It is procedural. Every spacecraft launched from Earth carries with it traces of Earth's biosphere. No sterilization process is perfect. Some bacteria survive intense radiation, extreme cold, vacuum, and chemical assault. Tardigrades, microscopic animals found in Earth's mosses and sediments, have demonstrated the ability to survive the conditions of space itself. Clean rooms used to assemble spacecraft routinely contain microbial life despite rigorous controls.

History offers uncomfortable reminders. Apollo astronauts returned to Earth carrying bacteria that had survived the lunar environment inside their spacecraft. Surveys of supposedly sterile facilities consistently detect resilient organisms. Reports suggesting that Streptococcus mitis survived on the Moon remain debated, but the fact that the debate exists at all underscores the uncertainty involved.

Europa raises the stakes dramatically. If a probe were to penetrate the ice shell, intentionally or accidentally, and introduce Earth microbes into the ocean, the consequences could not be undone. Those microbes might die quickly, unable to adapt to Europa's chemistry and pressure. Or they might survive, finding an environment free of competition, where nutrients and energy sources were available. In that case, Earth life could proliferate, displacing or erasing any native biosphere that existed there.

The loss would be absolute. Once contamination occurred, there would be no way to determine which organisms were native and which were invaders. The scientific value of the ocean would be permanently compromised. If native life were destroyed, humanity would have committed an act with no precedent in history, the eradication of an entire alien biosphere before it was even understood.

Backward contamination presents a different but related concern. If material from Europa were returned to Earth, could alien organisms pose a risk to Earth's biosphere or human health. Most scientists consider catastrophic scenarios unlikely, but uncertainty remains. Earth's immune systems evolved in isolation from alien biology. Caution is not paranoia when the unknown is truly unknown.

Current planetary protection protocols struggle to meet this challenge. The Committee on Space Research classifies missions based on contamination risk, but enforcement mechanisms are weak. Europa flybys fall under less restrictive categories than landers. Commercial missions often operate in regulatory gray zones. There is no global authority with the power to halt a mission deemed too risky.

The gap between risk and governance is widening rapidly.

The Ethical Bomb

The discovery of water and potential life beneath Europa's ice detonates a moral dilemma unlike any humanity has faced. It is not a choice between good and evil, but between competing goods that cannot be easily reconciled.

Imagine that microbial life in Europa's ocean is confirmed beyond reasonable doubt. Imagine also that the same ocean contains vast quantities of water that could be split into hydrogen and oxygen to fuel spacecraft. That water could enable deep space missions, asteroid deflection, planetary defense, and perhaps even the long term survival of humanity beyond Earth. Extracting it would almost certainly damage the ecosystem and could extinguish life that took billions of years to emerge.

Do we exploit the resource. Do we protect the life. Can we do both.

Anthropocentric ethics prioritize human needs above all else. From this perspective, humanity's survival and flourishing outweigh the interests of microbes that lack consciousness or suffering. Humans routinely kill trillions of bacteria to save a single life. Europa's microbes, proponents argue, are not fundamentally different. Resources that could save human lives should not be left unused.

The counterargument is not sentimental. Earth's microbes evolved alongside us within a shared biosphere. Europa's life would represent an independent origin, a separate evolutionary experiment with intrinsic value precisely because it is not ours. Destroying it would not be routine sterilization. It would be a choice to erase an entire branch of life from the universe.

Biocentric perspectives hold that all life has intrinsic value regardless of complexity. From this view, the rarity of life elevates its moral status. Europa's biosphere, if it exists, would be priceless and irreplaceable. Convenience and even utility would not justify its destruction.

Critics of biocentrism note that taken to extremes, it paralyzes action. Humans cannot live without killing microbes. Drawing a moral boundary becomes difficult.

Ecocentric ethics offer a different lens, focusing on systems rather than individuals. The question becomes whether it is possible to meet human needs while preserving the integrity of the ecosystem. This framework emphasizes stewardship, restraint, and long term thinking rather than absolute prohibition or unchecked exploitation.

Utilitarian approaches attempt calculation, weighing total benefits against total harms. Yet this approach falters when confronted with alien life. How does one quantify the value of a second genesis. Whose interests count, present humans, future humans, or unknown organisms.

Deep ecology rejects instrumental thinking altogether, arguing that natural systems possess value independent of human use. From this perspective, Europa's ocean deserves protection not because it benefits humanity, but because it exists.

No framework offers an easy answer. The moral status of microbes complicates matters further. They are invisible, lack charisma, and do not suffer in ways humans recognize. Yet their significance is unparalleled. They would represent humanity's first encounter with life beyond Earth.

The question of authority looms over every argument. In practice, decisions will be made by whoever arrives first. Space agencies, corporations, and governments will act based on their own incentives. Yet the discovery of alien life would belong to all of humanity, including future generations who will live with the consequences.

No decision making framework exists that matches the scale of the stakes. The timeline is short. Missions are being planned now. Humanity is unprepared.

The Corporate Race

Water is not only a symbol of life. It is the foundation of the space economy. Split into hydrogen and oxygen, water becomes rocket propellant. Launching fuel from Earth is prohibitively expensive. Accessing water in space would transform exploration from a series of isolated missions into a sustained presence.

Europa's ocean represents an almost unimaginable reservoir. Its location in the outer solar system makes it attractive for deep space missions. In purely economic terms, it could become the most valuable resource beyond Earth.

Space agencies frame their missions as scientific, but resource potential is quietly acknowledged. Private companies, though unable to reach Europa in the near term, position themselves strategically. The pattern is familiar from Earth's history. Exploration leads to assessment. Assessment leads to utilization. Utilization leads to exploitation.

Corporate logic is difficult to resist. Companies are obligated to maximize returns. First mover advantage matters. Competition creates pressure to act quickly. Environmental protection is framed as delay. Governments, increasingly reliant on private investment, struggle to regulate the very entities that enable their ambitions.

Debate intensifies. An astrobiologist compares drilling into Europa's ice to burning the Library of Alexandria for warmth. An engineer responds that water could fuel missions that prevent human extinction. Both arguments contain truth. Neither offers resolution.

The danger lies not in greed, but in momentum.

The Line We Have Not Crossed Yet

Europa's ocean remains sealed beneath ice, protected for now by distance and difficulty. That protection will not last forever. Within decades, humanity will face a decision that cannot be undone.

One future sees Europa penetrated, sampled aggressively, and exploited. No life is found, or it is destroyed before discovery. The pattern spreads outward to Enceladus, Titan, and beyond. Another future sees Europa studied carefully from orbit, life protected, and alternative resources pursued elsewhere. Humanity demonstrates that it has learned from Earth.

The choice is coming faster than most realize.

Water lies beneath distant moons. Ice rests on Mars. Minerals drift through asteroids. The resources are there, and the race to claim them is accelerating.

The question is no longer whether humanity can reach them. It is whether humanity can restrain itself once it does.

And who gets to decide.

8

The Asteroid Gold Rush

Sixteen Psyche. A metallic asteroid orbiting between Mars and Jupiter. Two hundred twenty six kilometers across. Composed almost entirely of iron, nickel, and precious metals. Estimated value ten thousand quadrillion dollars. Written out, that is ten quintillion dollars, more than the entire global economy by a factor so large it defies intuition. It is a mountain of metal drifting silently through space, a remnant of a shattered protoplanet, untouched for four and a half billion years.

It is just floating there. Waiting. And everyone wants a piece.

In 2020, NASA approved a mission to visit Psyche. Officially, it was a science mission, designed to study a metallic asteroid up close for the first time and learn how planetary cores form. The spacecraft launched in 2023 and is scheduled to arrive in 2029. By the time it gets there, a dozen private companies will have internal plans detailing how to extract wealth from objects like it. The question is no longer whether humanity will mine asteroids. The question is how soon, and at what cost.

The language surrounding asteroid mining is seductive. Limitless resources. Infinite growth. Environmental salvation through off-world extraction. It sounds like a clean solution to the mess we made on Earth. Yet gold rushes have a history, and they rarely unfold the way their architects imagine. They begin with optimism and end with scars.

This one is already underway.

The Treasure Map

To understand the intensity of the asteroid gold rush, it helps to understand what is actually out there. Popular imagination often pictures the asteroid belt as a dense field of tumbling rocks, but the reality is more subtle and, in some ways, more dangerous. The asteroid belt lies between the orbits of Mars and Jupiter, spanning roughly one to three astronomical units from the Sun. It contains millions of objects, yet the total mass of all those asteroids combined is only about four percent of the Moon's mass. What makes them valuable is not their collective size, but their composition and accessibility.

Most asteroids are leftovers from the early solar system, material that never coalesced into planets due to Jupiter's gravitational influence. They are chemically primitive, preserving the building blocks from which planets formed. Some are rocky, some are rich in water and organics, and some are essentially floating metal deposits larger than any mine on Earth.

While the main belt captures imagination, the real focus of near-term mining efforts lies much closer to home. Near-Earth asteroids are objects whose orbits bring them within about one point three astronomical units of the Sun. Roughly thirty thousand have been cataloged so far, and astronomers estimate that perhaps a million more exist. Some are only meters across, others stretch for kilometers. Crucially, many of them require less energy to reach than the Moon or Mars. From an engineering standpoint, they are low hanging fruit.

Asteroids fall into three broad categories. Carbonaceous asteroids, known as C-types, make up roughly three quarters of the population. They contain water locked in hydrated minerals, along with organic compounds, clays, and silicates. Their economic value in terms of precious metals is modest, but their strategic value is immense. Water can be split into hydrogen and oxygen for rocket fuel. It can support life. It can enable a permanent space presence. Bennu, the target of NASA's OSIRIS-REx mission, is a carbonaceous asteroid about five hundred meters wide, with an estimated water content of around ten percent. That represents billions of liters of water drifting through space.

Silicaceous asteroids, or S-types, account for about seventeen percent of known asteroids. They are composed primarily of iron, nickel, and magnesium silicates, with smaller amounts of precious metals. These bodies are attractive as sources of construction material. Iron and nickel can be used to build structures in space, reducing the need to launch heavy materials from Earth. The asteroid Itokawa, visited by Japan's Hayabusa mission, is a classic example, a small, irregular object rich in metal and stone.

The rarest and most coveted are M-type asteroids. These metallic bodies, making up perhaps eight percent of asteroids, are composed mostly of iron and nickel, with significant concentrations of platinum group metals such as platinum, palladium, rhodium, iridium, and osmium. Even a small M-type asteroid could contain more platinum than has ever been mined in human history. Sixteen Psyche is the extreme case, a world-sized lump of metal that may represent the exposed core of a failed planet. Its iron content alone would dwarf all terrestrial reserves. Its precious metals are beyond meaningful calculation.

NASA's Center for Near-Earth Object Studies maintains a growing database of asteroids, cataloging their orbits, compositions, and accessibility. Using spectroscopy and radar, scientists can estimate what these objects are made of and how difficult they are to reach. Lists of top mining targets already circulate within industry circles. Asteroids like Eros, Amun, Nereus, and

Ryugu are discussed not as celestial bodies, but as balance sheets.

The numbers are staggering. Humanity has mined roughly two hundred thousand tons of gold in all of history, worth about twelve trillion dollars at current prices. A single one kilometer metallic asteroid could contain two hundred million tons of platinum group metals. At current market prices, that translates to several quadrillion dollars. Introducing that volume of material into the economy would collapse prices, upend industries, and reshape global power. Scarcity, as we understand it, would evaporate.

So why now. The answer lies in convergence. Reusable rockets have driven launch costs down dramatically. Robotics and artificial intelligence enable autonomous operations far from Earth. Advances in materials science and additive manufacturing make in-space processing plausible. Meanwhile, billions of dollars in private capital search for the next frontier. Asteroid mining sits at the intersection of technological feasibility and financial ambition.

The treasure map is no longer speculative. It is being actively drawn.

The Technology

Asteroid mining is often dismissed as fantasy, but the foundational technologies already exist. Over the past two decades, space agencies have quietly demonstrated the most difficult steps required to make it real.

NASA's OSIRIS-REx mission rendezvoused with the asteroid Bennu, mapped it in exquisite detail, touched its surface, collected material, and returned that sample to Earth in 2023. Japan's Hayabusa and Hayabusa2 missions accomplished similar feats at Itokawa and Ryugu, returning samples despite mechanical failures and unexpected terrain. These missions proved that spacecraft can navigate to small, irregular bodies, operate in microgravity, interact with their surfaces, and bring material home.

The next phase is prospecting. NASA's Psyche mission will be the first to visit a metallic asteroid, studying its composition, magnetic field, and structure. The Lucy mission is surveying Trojan asteroids near Jupiter, while missions like NEA Scout and DART demonstrate navigation and orbital manipulation. DART, in particular, showed that humans can deliberately alter an asteroid's trajectory, a capability with profound implications.

Private companies have come and gone, but their ideas persist. Concepts range from capturing small asteroids and bringing them into lunar orbit to processing materials directly on site. One proposed method involves enclosing part of an asteroid in a reflective bag and using concentrated sunlight to vaporize water and volatiles, which can then be collected. Others envision drilling, surface scraping, or magnetic separation for metallic bodies.

Each approach carries tradeoffs. Capturing asteroids simplifies processing but introduces political and safety concerns, as redirecting even a small object carries impact risk. In situ processing avoids transport costs but demands highly reliable robotics operating far from human intervention. Communication delays mean machines must make decisions autonomously, and repairs are often impossible.

Processing materials in space presents additional challenges. Traditional mining relies on gravity, water, and atmosphere, all absent or limited in space. New techniques such as solar furnaces, electromagnetic separation, and microgravity electrolysis are being developed. Water becomes both the resource and the bottleneck, since many processes require it. This creates a logic loop where water rich asteroids must be mined first to enable mining of others.

Infrastructure is the hidden cost. Asteroid mining requires launch vehicles, orbital depots, transit craft, autonomous miners, processing facilities, and either return vehicles or in-space manufacturing hubs. None of this exists at scale yet, but pieces are falling into place. Governments and private investors have already spent more than ten billion dollars on relevant technologies.

Timelines vary, but most serious projections place small scale mining in the 2030s or 2040s, with industrial operations following later. This is not distant speculation. It is a medium-term plan unfolding now.

The Environmental Unknowns

Asteroids are not just resources. They are time capsules. Each one preserves a unique record of the early solar system, material that predates Earth itself. Unlike planets, most asteroids were never melted or differentiated. Their chemical and isotopic signatures offer clues to how planets formed, how water arrived on Earth, and how organic molecules were distributed through space.

Sample return missions bring back grams of material. Scientists study those grains for decades. Mining operations would process millions of tons, obliterating information that can never be recovered. Once an asteroid is mined, its scientific record is destroyed. There is no way to UN-mine a celestial body.

Mining also creates debris. Broken equipment, processing waste, and fragmented material become long lived hazards. In Earth orbit, debris cascades can render regions unusable. While mining in the main belt poses little immediate risk, operations involving near-Earth asteroids intersect with heavily trafficked space lanes. Failed equipment and unintended fragments could persist for centuries.

Orbital dynamics introduce further uncertainty. Mining changes mass distribution and momentum. Even small alterations can compound over decades. A trajectory that is safe today might not remain so after intervention. Planetary defense depends on precise predictions. Mining introduces noise into a system where precision matters.

Contamination adds another layer of risk. Asteroids contain organic compounds that may resemble the precursors of life. Introducing Earth microbes compromises the ability to study those materials. Returning samples to Earth carries minimal but nonzero risk, one that currently falls under company self assessment rather than independent review.

At present, the entities assessing these risks are the same ones poised to profit. There is no binding requirement for environmental impact analysis. No international scientific review board. No precautionary pause.

The unknowns are vast, and the incentives to ignore them are strong.

The Legal Vacuum

The legal framework governing space was written for a different era. The Outer Space Treaty of 1967 declares that space is the province of all humankind and forbids national appropriation. It says little about resource extraction. Companies interpret this silence as permission.

The analogy often used is fishing in international waters. No one owns the ocean, but you own the fish you catch. The United States codified this interpretation in its 2015 Space Act, granting citizens rights to resources they extract in space. Luxembourg followed with similar legislation in 2017, investing heavily to attract space mining companies. Other nations are drafting their own laws.

There is no international consensus. Some countries contest this interpretation. No court has jurisdiction over disputes in space. No enforcement mechanism exists. Benefit sharing, nominally required by treaty, remains theoretical.

History offers familiar parallels. First come, first served regimes favor the wealthy and powerful. Environmental costs are externalized. Inequality deepens. Once established, these systems are difficult to reverse.

Elena Rodriguez, founder of Astro Mine Industries, sees opportunity. A former aerospace engineer, she believes asteroid mining will unlock abundance and reduce pressure on Earth. To her, environmental concerns applied to space are misplaced. Space is vast. Asteroids are inert. There is nothing to harm.

Dr. James Chen, a planetary scientist, sees loss. Each asteroid mined is a library burned. He argues for protected bodies, for study before extraction, for governance before profit. He knows the momentum is against him.

Their debate reflects a broader conflict. Neither position is absurd. Both carry weight. The danger lies not in choosing wrong, but in choosing quickly without understanding the consequences.

History, Repeating

The asteroid gold rush has begun. Companies are forming. Capital is flowing. Technology is advancing. Legal frameworks are hardening around permissive interpretations. Within two decades, mining will likely commence.

History offers warnings. Gold rushes enrich a few and scar landscapes for generations. Rivers destroyed in nineteenth century California remain damaged today. The Klondike's wealth was fleeting. Its scars were not.

Asteroids may seem different. They are distant. They appear lifeless. Yet the patterns are the same. Excitement. Rush. Extraction. Regret.

There are more than ten thousand near-Earth asteroids considered economically valuable. None are protected by international law. The total estimated value runs into the quadrillions. The window to establish governance is closing fast.

Once the machinery of extraction is in motion, stopping it becomes nearly impossible.

Asteroids are the frontier everyone is watching. But while we look outward toward distant rocks, we are already creating a crisis much closer to home. Right above our heads, in the orbital space around Earth, debris is accumulating. And that crisis threatens not abundance, but access itself.

9

Orbital Graveyard

"Houstoun, we have a problem."

The call came at 3:47 a.m. Eastern Standard Time, cutting through the quiet hum of mission control like a dropped plate. The International Space Station had just executed an emergency maneuver. Seven astronauts from five nations were asleep or floating through routine tasks when the station's guidance system fired its thrusters, nudging the entire structure a few hundred meters sideways in its orbit. The reason was a piece of debris that had not been there the last time the orbital path was checked.

Later analysis traced the fragment to a Chinese weather satellite destroyed fifteen years earlier during an anti-satellite test. The debris was no bigger than a marble. It passed within three hundred meters of the station.

Three hundred meters sounds comfortable on Earth. In orbit, at a closing speed of more than seventeen thousand miles per hour, it is the difference between survival and annihilation. That fragment carried the kinetic energy of a hand grenade. Had it struck the station, the result would have been instant depressurization, catastrophic structural failure, and the likely deaths of everyone on board. Humanity's most complex machine would have become just another cloud of shrapnel, adding thousands of new fragments to an already saturated orbital environment.

This was not an extraordinary event. The International Space Station maneuvers to avoid debris several times each year. Since its first crew arrived in 1999, it has performed more than thirty avoidance maneuvers, and the time between them keeps shrinking. Astronauts now train for debris threats as routinely as they train for fire or pressure leaks. The risk has become background noise, ever present, rarely discussed outside specialist circles.

We are building a prison around our own planet. It is invisible, fast moving, and almost impossible to clean up. If it closes, it may close for generations.

The Space Junk Crisis

Space looks empty when we look up at night, but the region around Earth has become one of the most densely polluted environments humans have ever created. As of 2024, more than thirty six thousand objects larger than ten centimeters are actively tracked by the United States Space Surveillance Network. These are only the pieces big enough to be seen and cataloged. Between one and ten centimeters, the estimated number jumps to around one million. Below one centimeter, there may be more than one hundred million fragments, including paint flecks, metal shards, and pulverized insulation.

At orbital velocities, size matters less than speed. In low Earth orbit, objects travel at roughly seven to eight kilometers per second. When two objects collide head on, relative velocities can reach fifteen kilometers per second. Kinetic energy increases with the square of velocity, which means that even a tiny object becomes a weapon. A one centimeter fragment carries the energy of a bowling ball thrown at highway speed. A ten centimeter object delivers the energy of several kilograms of explosive. No spacecraft shielding can protect against impacts at those energies. Avoidance is the only defense.

Orbital debris is not evenly distributed. Low Earth orbit, extending from about one hundred sixty kilometers to two thousand kilometers above the surface, is the most crowded region. It is also the most valuable. Earth observation satellites, crewed stations, scientific missions, and mega-constellations all compete for space here. Atmospheric drag eventually pulls objects down, but decay times vary widely. At four hundred kilometers, debris may fall within years. At eight hundred to one thousand kilometers, it can persist for centuries.

Medium Earth orbit, home to navigation systems like GPS, is less congested but more persistent. Debris here can remain aloft for thousands of years. Geostationary orbit, nearly thirty six thousand kilometers above Earth, is the most valuable real estate of all. Satellites here remain fixed over a point on the planet, making them indispensable for communications and weather monitoring. Debris at this altitude effectively never comes down. It becomes a permanent part of Earth's orbital environment.

The sources of this debris are many. Spent rocket stages account for the majority of large objects, abandoned after delivering their payloads. Mission related debris, such as separation bolts and protective covers, adds thousands more. Satellites themselves often break apart when batteries explode or propellant tanks rupture years after launch. More than two hundred ninety on-orbit breakups have been recorded.

Then there are the deliberate acts. Anti-satellite weapons tests have created some of the worst debris events in history. In 2007, China destroyed its Fengyun-1C weather satellite, producing more than three thousand trackable fragments and hundreds of thousands of smaller ones. In 2021, Russia destroyed Cosmos 1408, creating fifteen hundred tracked pieces that immediately endangered the International Space Station. These tests were widely condemned, yet they continue, justified as demonstrations of capability and deterrence.

Accidental collisions add another layer of chaos. In 2009, the operational Iridium 33 satellite collided with the defunct Russian Cosmos 2251 at nearly twelve kilometers per second. Both were obliterated. More than two thousand fragments large enough to track were created, along with hundreds of thousands of smaller pieces. That debris cloud is still spreading, intersecting with other orbits and raising collision risks for unrelated spacecraft.

The growth curve tells a worrying story. For decades, debris accumulated slowly. Then the curve bent upward. The Fengyun test in 2007 marked a sharp spike. Since 2020, the rise has become exponential, driven by mega-constellations that plan to place tens of thousands of satellites into low Earth orbit. Each launch adds capability, but also risk. Each collision multiplies the danger.

The International Space Station lives inside this environment. Its Whipple shields can absorb impacts from particles up to about one centimeter, but anything larger is lethal. When collision probabilities exceed one in ten thousand, avoidance maneuvers are triggered. Each maneuver burns fuel, disrupts experiments, and shortens the station's lifespan. Astronauts now regularly shelter in their return vehicles during high risk periods, ready to evacuate at a moment's notice.

The psychological toll is real. Debris cannot be seen. There is no warning sound. There is only probability. Living in orbit has become a form of cosmic roulette.

Kessler Syndrome Explained

In 1978, NASA astrophysicist Donald Kessler published a paper that few outside the field noticed at the time. In it, he described a scenario in which the density of objects in Earth orbit reached a point where collisions would become self sustaining. Each collision would create debris that caused further collisions, leading to a cascading effect that could render entire orbital regions unusable.

The mechanism is straightforward. Two objects collide and shatter, producing thousands of fragments. Each fragment becomes a high velocity projectile capable of destroying other objects. As debris density increases, collision rates rise. Eventually, debris is created faster than it can decay out of orbit. At that point, the process feeds on itself.

Kessler's insight was that this threshold exists, and that crossing it would fundamentally change Earth's orbital environment. Below the threshold, debris can be managed. Above it, no amount of caution prevents escalation.

The 2009 Iridium-Cosmos collision provided the first real-world proof. It was not an act of war or negligence. It was a statistical inevitability. Two objects occupied the same space at the same time, and physics did the rest. The debris created that day continues to threaten spacecraft fifteen years later.

Current assessments are sobering. Both NASA and the European Space Agency acknowledge that some orbital shells may already be in the Kessler regime. This does not mean an immediate catastrophe, but it does mean that collision rates are likely to increase even if no new objects are launched. The system has momentum.

If a full cascade occurs in low Earth orbit, the consequences would be severe. Satellites would be destroyed faster than they could be replaced. Launching new spacecraft would become prohibitively dangerous, as vehicles would have to pass through dense debris fields during ascent. Weather forecasting, global communications, navigation, Earth observation, and much of modern life would be disrupted.

The loss would not be temporary. Debris at higher altitudes could persist for centuries. Humanity could find itself effectively trapped on Earth, unable to safely access space during the period when space access is most critical to planetary monitoring and defense.

The cruel irony is that the same orbital region most vulnerable to cascade is also the most useful. Low Earth orbit is where we put the systems that help us understand and manage our planet. Losing it would be a self inflicted wound of staggering proportions.

Why Cleanup Is So Hard

Cleaning up space debris sounds simple until one confronts the physics. Debris travels too fast to be easily captured. Objects are too numerous to track individually. There is no gravity to pull fragments together, no atmosphere to slow them down. Any cleanup system must rendezvous with targets moving at kilometers per second, grab them without creating more debris, and dispose of them safely.

Proposals exist. Nets, harpoons, robotic arms, drag sails, lasers that nudge debris into lower orbits. Each works only for certain sizes and conditions. None address the smallest fragments, which collectively pose enormous risk. Even removing a few large objects would require sustained, expensive international effort.

The greatest obstacle is incentive. No one owns debris. Cleaning it up generates no direct profit. The benefits are shared by all, while the costs are borne by the cleaner. This is the classic tragedy of the commons, playing out at orbital speed.

Prevention is far easier than cleanup. Designing satellites to deorbit at end of life. Passivating fuel tanks to prevent explosions. Avoiding destructive tests. Coordinating traffic. All are technically feasible. None are universally enforced.

The window for prevention is narrowing. Every year of delay increases the baseline risk. Once a cascade accelerates, options disappear.

The Governance Failure

There is no global traffic control system for space. There is no binding international law that limits debris creation or mandates cleanup. The Outer Space Treaty was written when there were only a handful of satellites in orbit. It does not address congestion, liability for debris, or long-term environmental stewardship.

National regulations exist, but they are inconsistent and often favor domestic industries. Mega-constellation operators promise responsible behavior, but commercial pressure rewards speed and scale. Anti-satellite tests continue despite universal condemnation.

We have seen this pattern before. When a resource is shared, valuable, and weakly governed, exploitation outpaces caution. By the time consequences are undeniable, they are often irreversible.

The Closing Door

Orbital space is not infinite. It is a narrow shell around a fragile planet. We have treated it as empty because we could not see the damage accumulating. Now we can measure it, model it, and feel it in emergency maneuvers and rising risk probabilities.

The orbital graveyard is not a future problem. It is a present one. We are at the inflection point Donald Kessler warned about nearly half a century ago. The choices we make in the next decade will determine whether Earth remains a space faring world or retreats into orbital isolation.

The tragedy is that we know what to do. The failure is that we have not done it.

Asteroids promised limitless wealth. Orbits promised effortless access to space. Both are constrained by physics and ethics we have been slow to acknowledge. As we turn to the final chapters of this story, the question is no longer whether space needs environmental protection.

The question is whether we are willing to act before the door closes behind us.

10

Are We Alone? (And Does It Matter?)

The most profound question humanity has ever asked might have a disturbing answer.

We were alone until we killed them.

Before we even knew they existed.

Imagine a scene that no longer belongs to science fiction. It is the year 2035. A rover on Mars, equipped with a drill designed to reach beneath the planet's frozen crust, penetrates a pocket of subsurface ice. Instruments register complex organic chemistry. Metabolic activity. Replication. The signal is unmistakable. Microbial life. Martian life.

The announcement is immediate and electric. The discovery of the millennium. Proof that biology is not unique to Earth. A second genesis, billions of years old, evolving in isolation under a pale red sky. Humanity answers a question it has asked since the first humans looked up at the stars and wondered whether the darkness looked back.

Then comes the second announcement, quieter and slower. The microbes are dying.

Genetic analysis reveals contamination. Earth bacteria, carried inadvertently by the rover despite sterilization protocols, have infiltrated the sample site. Within months, the native organisms collapse. They cannot compete. The alien biosphere, already marginal in its extreme environment, is extinguished. Not by malice. By carelessness.

This is not a fantasy scenario. It is a realistic outcome based on current planetary protection practices, economic pressures, and technological trajectories. We are searching for life beyond Earth. We are very close to finding it. But we are not yet prepared to protect it.

And that failure would echo far beyond science.

The Search for Life

For most of human history, the question "Are we alone?" belonged to philosophy and religion. In the past half century, it has become an empirical problem. We have learned where to look, how to look, and what signs to search for. The universe has revealed itself as far more chemically fertile than we once imagined.

Mars: the near neighbor

Mars remains the most accessible and emotionally charged candidate for extraterrestrial life. Its surface tells a clear story. Ancient river valleys snake across its plains. Delta fans mark the mouths of long-dried lakes. Minerals formed only in the presence of neutral, drinkable water appear in abundance. For hundreds of millions of years, early Mars was warm, wet, and stable.

NASA's Perseverance rover, operating in Jezero Crater since 2021, is exploring what was once a lake fed by a river system comparable in scale to those on early Earth. The rover has drilled sedimentary rocks laid down underwater, discovering organic molecules embedded within them. These compounds do not prove life, but they are the raw materials from which life can arise.

Methane complicates the picture further. Curiosity has detected episodic methane spikes in the Martian atmosphere, sometimes reaching ten times background levels. The methane appears seasonally, responding to temperature changes. On Earth, the vast majority of methane is produced by biology. Mars has geological processes that could also produce it, but no explanation yet accounts for the observed variability.

The scientific consensus is careful. Mars was habitable. Life may have arisen. If it did, it could persist today in subsurface ice, deep aquifers, lava tubes, or other protected environments shielded from radiation and cold. Mars may not be dead. It may be quiet.

Ocean worlds: life beneath ice

Beyond Mars, the strongest candidates for existing life lie far from the Sun. Europa, Enceladus, Titan, and other icy moons harbor vast subsurface oceans kept liquid by tidal heating. Europa's ocean alone contains more water than all of Earth's oceans combined.

Enceladus, a small moon of Saturn, stunned scientists when the Cassini spacecraft flew through geysers erupting from its south pole. Those plumes contained water vapor, salts, complex organic molecules, and molecular hydrogen. The hydrogen is especially important. On Earth, it is a key energy source for microbial ecosystems surrounding hydrothermal vents, environments completely independent of sunlight. Cassini also detected silica grains consistent with hot water interacting with rock on the ocean floor.

This combination liquid water, organic chemistry, and chemical energy represents the same conditions under which life may have originated on Earth. If life can exist in the dark depths of Earth's oceans, fed by chemistry rather than sunlight, there is no obvious reason it could not exist beneath Enceladus's ice.

Europa shows similar promise. Its fractured ice shell, induced magnetic field, and possible water plumes all point to a global ocean in contact with a rocky mantle. Missions now en route will study it in unprecedented detail.

Titan and exotic possibilities

Titan expands the imagination further. It has lakes, rivers, and rain but not of water. Liquid methane and ethane flow across its surface under a nitrogen-rich sky. Beneath the ice, Titan also appears to harbor a water ocean. This raises the possibility of two separate environments where life could exist: a conventional water-based biosphere below, and an exotic methane-based chemistry above.

Such life would not resemble anything on Earth. Its biochemistry would be slower, colder, and radically different. But difference is not impossibility. Titan reminds us that our assumptions about life are shaped by our own example, not by universal law.

Beyond the solar system

The final blow to cosmic loneliness came with the discovery of exoplanets. Since the 1990s, astronomers have confirmed more than five thousand worlds orbiting other stars. Many lie in their star's habitable zone. Some are Earth-sized. Some appear to have atmospheres. A few already show tantalizing hints of water vapor and chemical disequilibrium.

The James Webb Space Telescope can now analyze the atmospheres of distant planets, searching for combinations of gases that should not coexist unless constantly replenished by life. Oxygen and methane together. Carbon dioxide with reduced compounds. These are subtle signals, but they are detectable.

The statistics are staggering. There are billions of planets in our galaxy alone. The building blocks of life are found in meteorites and interstellar clouds. Life emerged on Earth quickly, as soon as conditions allowed. The conservative view remains cautious. Our sample size is one. But the optimistic view is increasingly difficult to dismiss.

Life seems not miraculous, but persistent.

What Counts as Life?

Before we can decide how to treat extraterrestrial life, we must confront an uncomfortable fact. We do not even agree on what life is.

Biology textbooks offer familiar criteria: metabolism, growth, reproduction, responsiveness, homeostasis, evolution. Yet edge cases abound. Viruses replicate and evolve but lack metabolism. Prions replicate without cells. Mules are sterile yet undeniably alive. Fire grows, spreads, and consumes energy but is not life.

NASA's working definition"a self-sustaining chemical system capable of Darwinian evolution" is useful but incomplete. It assumes chemistry we recognize and mechanisms we understand. Alien life may violate our categories.

Earth itself has repeatedly embarrassed our assumptions. Extremophiles thrive in boiling water, acidic mine drainage, Antarctic ice, hyper saline lakes, and deep underground rock. Deinococcus radiodurans can survive radiation doses that would pulverize human DNA. Some microbes persist in a state of near-suspended animation for thousands of years.

These discoveries matter because they expand the range of environments where life might exist and the forms it might take. Alien life is unlikely to be charismatic. It will not wave, communicate, or even move in ways we recognize. It will probably be microbial.

And that creates an ethical problem.

The instinctive response is dismissal. We kill microbes constantly. Antibiotics save lives by destroying bacteria. Hand sanitizer annihilates trillions without moral concern. Why should alien microbes be different?

There are several answers.

First, alien life would represent the only independent origin of biology we have ever encountered. Its scientific value is incalculable. It could reveal whether life is inevitable or contingent, whether DNA is universal or parochial, whether evolution follows predictable paths. Destroying it would erase knowledge no technology could recreate.

Second, we would not be killing isolated organisms but entire biospheres. Even microbial ecosystems involve interactions, feedback loops, and long evolutionary histories. We would be ending a story billions of years in the making.

Third, every complex organism on Earth descended from microbes. Had early life been extinguished, nothing else would have followed. Protecting microbial life is not sentimentality. It is recognition of potential.

Finally, there is precedent. How we treat the first life we find beyond Earth will define our cosmic ethics. It will signal whether we see the universe as a partner to understand or a warehouse to strip.

The Contamination Threat

The greatest danger to extraterrestrial life does not come from malice. It comes from microbes we cannot see and protocols that cannot guarantee sterility.

No spacecraft is sterile. Clean rooms reduce contamination but do not eliminate it. Even after chemical treatment, heat sterilization, and radiation exposure, hardy spores survive. Some bacteria are remarkably resilient. Tardigrades, microscopic animals, can survive vacuum, radiation, and extreme temperatures.

We know this because we have tested it. And because we have failed.

During the Apollo program, astronauts retrieved parts of the Surveyor 3 lander that had sat on the Moon for two and a half years. Living bacteria were reportedly found, though later studies questioned whether contamination occurred during retrieval. Regardless, the episode shattered assumptions about survivability.

More recently, the Beresheet lunar lander crashed in 2019, scattering dormant tardigrades across the Moon's surface. They are probably inactive, but their presence illustrates the ease with which we contaminate other worlds.

Mars missions face far greater risks. Modern rovers are not sterilized to Viking-era standards because of cost and complexity. Perseverance drills into rocks. Future missions will drill into ice, the very places life is most likely to persist.

Once contamination occurs, reversal is impossible. Introduced organisms spread. Some lie dormant. Some adapt. On Earth, invasive species have devastated ecosystems despite intense eradication efforts. A planet offers no such control.

Backward contamination is less likely but not negligible. Mars Sample Return missions plan extreme containment measures. But unknown biology carries unknown risks. The probability of harm may be low, but the consequences could be severe.

Planetary protection policies exist, but they rely largely on self-regulation. Commercial missions often operate under looser standards. There is no enforcement mechanism. No global authority. And immense pressure to move quickly.

Ethical Frameworks

How, then, should we act?

Anthropocentrism places human interests above all else. From this perspective, alien microbes have little moral standing. Human survival, expansion, and prosperity matter more. If resources on Mars enable human flourishing, they should be used.

Biocentrism argues that all life has intrinsic value. Alien microbes, precisely because they are rare and independent, deserve protection. Destroying them for convenience is morally indefensible.

Ecocentrism shifts focus to systems rather than individuals. An entire Martian biosphere has value even if individual microbes do not. Sustainable interaction might be possible, but only after thorough understanding.

The precautionary principle urges restraint under uncertainty. When harm is irreversible and knowledge incomplete, the burden of proof lies with those who wish to act.

Utilitarianism attempts calculation. Benefits versus costs. Knowledge versus resources. But some values resist quantification. What is the worth of a second genesis? How do we price the answer to a question humanity has asked for millennia?

Indigenous perspectives add another dimension. Many traditions emphasize relationship, reciprocity, and long-term thinking. Humans are not conquerors of nature but participants within it. Applied to space, this suggests humility. Respect for what we do not yet understand.

The Question Beneath the Question

Are we alone?

We do not yet know. But we are closer than ever. Drills are penetrating ice. Probes are flying through plumes. Telescopes are reading alien atmospheres. Within decades, perhaps within years, we may have an answer.

When we do, the more important question will follow immediately.

What do we do with the answer?

There are futures in which we discover life and protect it. We learn patiently, non-invasively, preserving places that teach us who we are and where we fit in the universe. There are futures in which we discover life and destroy it, repeating patterns of exploitation perfected on Earth. And there are futures in which we never know what we lost, because contamination erased the evidence before discovery.

The window for choosing among these futures is narrow. Planetary protection budgets are tiny compared to exploration costs. Decisions are being made quietly, by few people, on timelines driven by competition rather than reflection.

Humanity has waited tens of thousands of years to ask whether we are alone. We may answer the question within the next twenty. It would be a bitter irony to learn that we were not and that we ensured we would be again.

This outcome is not inevitable. It requires only that we slow down slightly, think more carefully, and treat the possibility of life with the seriousness it deserves.

Because the moment we find life beyond Earth, the universe will no longer be empty.

And the way we respond will say more about us than about them.

11

The Rights of Planets

In 2017, New Zealand passed a law that quietly challenged one of the deepest assumptions of modern civilization. The Whanganui River, a long and winding waterway that flows from the volcanic interior of the North Island to the Tasman Sea, was granted legal person hood. Under this law, the river was no longer classified as property owned by the state or managed solely for human benefit. Instead, it was recognized as a legal entity with rights, duties, and interests of its own, represented in court by appointed guardians.

This decision did not suggest that the river possessed consciousness or intention in any human sense. It did not claim that water could think or that currents could deliberate. What it acknowledged was something far more disruptive. The river existed as a whole system, older than any government, shaped by processes that could not be reduced to economic value. Harm to the river was not merely damage to a resource but injury to an entity whose continuity mattered.

The Whanganui law emerged from generations of Māori resistance to colonial frameworks that treated land and water as commodities. For the Whanganui iwi, the river was an ancestor rather than an object, a relationship rather than a possession. The legal system did not fully adopt that worldview, but it made space for it. It accepted that law could evolve to recognize forms of value that were not transactional.

Once that door was opened, it could not easily be closed.

As humanity extends its reach beyond Earth, the same foundational question reappears on a vastly larger scale. If a river can have rights, can a planet.

Legal Person hood for Nature

The recognition of legal rights for natural entities did not begin with Whanganui, nor did it end there. In 2008, Ecuador rewrote its constitution to include explicit Rights of Nature, recognizing that ecosystems possess the right to exist, persist, maintain themselves, and regenerate their vital cycles. This was not framed as environmental regulation but as constitutional principle. Nature itself became a rights holder, capable of being represented in court by citizens acting on its behalf.

This approach marked a departure from traditional environmental law, which typically balances economic development against environmental protection through cost benefit analysis. Under a rights based framework, destruction is not justified simply because it is profitable or efficient. The burden shifts. Those who would harm an ecosystem must justify why its integrity should be compromised at all.

New Zealand extended this logic further with Te Urewera, a vast forested region that was removed from state ownership and redefined as a legal entity. Te Urewera could not be sold, subdivided, or exploited as property. Guardians were tasked with acting in its interest, not as managers of resources but as stewards of a living whole. The forest was recognized as something that existed for its own sake as well as for the people who depended on it.

Similar ideas have appeared elsewhere, though often through court rulings rather than legislation. In Colombia, the Atrato River was recognized as a subject of rights following decades of environmental devastation linked to mining. In India, courts have at times recognized rivers as legal persons, though implementation has been uneven and contested. In the United States, some municipalities have passed ordinances recognizing rights of nature, even when those ordinances conflict with higher levels of law.

Taken together, these developments represent a slow but significant shift in legal philosophy. Nature is no longer seen exclusively as backdrop or inventory. It is increasingly recognized as participant.

This matters for space because existing space law remains rooted in an older worldview that treats celestial bodies as inert territory awaiting use. The Outer Space Treaty emphasizes non appropriation and peaceful exploration, but it says little about environmental harm and nothing about intrinsic value. It assumes that damage to space environments is either negligible or acceptable in pursuit of progress.

That assumption was formed when space activity was rare, expensive, and symbolic. It no longer reflects reality.

Today, humanity is preparing to drill into Martian ice, extract water from the Moon, mine asteroids for metals, and deploy massive orbital infrastructures that permanently alter near Earth space. These actions do not occur in environments that heal themselves. There is no erosion to erase scars, no biosphere to recycle waste, no atmospheric mixing to dilute contamination. Harm in space is durable, often irreversible, and frequently invisible until it is too late to undo.

The Rights of Nature movement provides a legal precedent for confronting this reality. It demonstrates that law can evolve to protect entities not because they are alive in a human sense, but because their integrity matters across time.

Do Lifeless Worlds Have Rights

The most persistent objection to planetary rights is deceptively simple. Rights, critics argue, belong to beings that can have interests, and interests require life or consciousness. Rivers support ecosystems. Forests host biodiversity. Mars appears lifeless. An asteroid is rock and metal. To grant rights to such bodies seems to stretch the concept beyond coherence.

Supporters of planetary rights respond by re framing what rights protect. Rights are not rewards for intelligence or suffering. They are tools societies use to prevent irreversible harm to things they deem valuable beyond immediate utility.

One argument centers on intrinsic value. A planet is not merely a pile of resources but a coherent system shaped by billions of years of geological and chemical processes. Mars preserves evidence of planetary evolution that Earth has erased through tectonics and erosion. Its surface is a record of atmospheric collapse, climate transition, and perhaps biological emergence. Destroying that record is not neutral. It is a permanent loss of knowledge that cannot be reconstructed.

Scientific value strengthens this claim. Lifeless worlds are natural laboratories that no human technology can replicate. A pristine asteroid contains information about the early solar system that is unavailable anywhere else. Mining it does not simply extract metal. It obliterates context. The difference between sampling and industrial processing is the difference between reading a page and burning the book.

There is also aesthetic and cultural value, which law already recognizes in other contexts. National parks are protected not because trees have rights but because societies decide that some places should remain intact for future generations. No one argues that Yellowstone is conscious, yet its preservation is widely accepted as ethical. Celestial bodies, older and more fragile than any terrestrial landscape, arguably deserve at least as much consideration.

The precautionary principle further complicates claims of lifelessness. Mars may harbor subsurface microbes. Europa almost certainly hosts an ocean with chemical energy. Declaring a world expendable based on current ignorance risks irreversible harm. Once contaminated or transformed, uncertainty becomes permanent.

Opponents argue that extending rights to lifeless worlds risks paralyzing human progress. They point out that human survival and prosperity may depend on accessing space resources. Water from the Moon could support deep space missions. Asteroid metals could reduce environmentally destructive mining on Earth. Legal restrictions could slow innovation and entrench inequality by favoring nations that already control space infrastructure.

They also invoke Antarctica as a counterexample. The continent is protected through treaty without granting it legal person hood. Cooperation, not rights, has preserved it. This suggests that governance frameworks can exist without extending legal standing to landscapes.

Supporters counter that Antarctica demonstrates precisely why planetary protection is possible. It shows that humanity can collectively decide to restrain exploitation in favor of long term scientific and environmental interests. Rights are not the only mechanism, but they are among the strongest.

Ultimately, the debate is not about rocks. It is about whether humanity sees the cosmos as a commons to be stewarded or a warehouse to be emptied.

Indigenous Cosmologies

Long before legal systems debated rights for rivers or planets, many cultures understood land and sky as relational rather than inert. Indigenous cosmologies across the world describe mountains, rivers, and celestial bodies as relatives or ancestors embedded in moral order.

The conflict surrounding Mauna Kea in Hawaii illustrates this tension. For astronomers, the mountain offers unparalleled viewing conditions. For Native Hawaiians, it is sacred ground tied to origin stories and identity. The debate is often framed as science versus belief, but that framing obscures the deeper issue. It is a clash between a worldview that treats land as instrument and one that treats it as kin.

Similar perspectives extend beyond Earth. Indigenous leaders from Navajo and Hopi communities have expressed opposition to mining the Moon or Mars, arguing that these bodies hold spiritual significance regardless of human presence. In these cosmologies, celestial bodies are not empty. They participate in a living universe.

Western legal systems often dismiss such views as symbolic, yet they encode principles that modern environmentalism struggles to articulate. Long term thinking, restraint, reciprocity, and humility are not policy add ons but foundational assumptions.

Imagining indigenous leaders drafting space treaties is not a romantic exercise. It exposes the narrowness of current frameworks. What if space law prioritized relationship over ownership, responsibility over entitlement, and continuity over growth.

Practical Implementation

Recognizing planetary rights does not require granting Mars a voice or consciousness. It requires institutions that represent planetary interests alongside human ones.

One proposal discussed in international forums involves establishing a United Nations Planetary Rights Committee. This body would evaluate proposed activities on celestial bodies and designate zones for protection, limited use, or open activity. Some regions would be preserved as scientific sanctuaries. Others could be used under strict environmental constraints.

Such a system would mirror marine protected areas on Earth, where extraction is limited or prohibited to preserve ecosystem function. Enforcement would rely on licensing, treaty obligations, shared monitoring, and liability frameworks rather than coercion.

These mechanisms already exist in international law. They are complex and imperfect, but they demonstrate that governance beyond national interest is possible.

In a committee room at the United Nations in the early 2030s, two lawyers sit across from one another. One argues that Mars should be recognized as a legal entity whose interests deserve representation before irreversible decisions are made. The other insists that granting rights to planets threatens innovation and human survival. Both speak carefully. Both believe they are protecting humanity's future.

What they are truly debating is not Mars. They are debating who we are becoming as a species.

Every generation defines value by choosing what it will protect. Rivers were once property. Forests were once limitless. The atmosphere was once thought infinite.

We were wrong each time.

The question is not whether Mars has rights. The question is whether we have the wisdom to grant them.

12

Lessons from Colonial History

Every time human societies believed they had discovered a new world, they told themselves a familiar story. They said the land was empty, or underused, or wasted on those who did not know how to extract its true value. They framed their arrival as progress rather than invasion, as opportunity rather than disruption, as destiny rather than choice. The result followed with almost mechanical consistency. Indigenous societies were displaced or destroyed. Ecosystems were transformed beyond recognition. Wealth flowed outward while poverty settled in. Centuries later, the consequences remained visible in borders, economies, and landscapes that still bear the scars of those first encounters.

This pattern was not the product of individual cruelty alone. It was systemic. It emerged from legal doctrines, economic incentives, religious justifications, and technological asymmetries that together formed a playbook. That playbook was refined across continents and centuries, tested repeatedly, and rarely questioned by those who benefited from it. Today, as humanity stands on the threshold of large scale activity beyond Earth, the language surrounding space exploration echoes these earlier moments with unsettling clarity.

The central question is not whether space exploration is wrong. Humans have always explored, and curiosity is not a sin. The question is whether we will repeat a historical pattern so well documented that ignorance can no longer be claimed as an excuse.

The Colonial Playbook

When Europeans arrived in the Americas at the end of the fifteenth century, they encountered societies that had adapted to their environments through generations of knowledge. The Taíno in the Caribbean practiced agriculture suited to island ecosystems, cultivated cassava and maize, managed fisheries, and organized themselves through complex social structures. Similar sophistication existed across the continents, from the agricultural terraces of the Andes to the urban centers of Mesoamerica. These societies were not static or idyllic, but they were functional, resilient, and deeply intertwined with their environments.

The arrival of European powers disrupted these systems almost immediately. Disease played a catastrophic role, but it did not act alone. Violence, forced labor, and deliberate restructuring of land use accelerated collapse. The encomienda system in Spanish colonies converted indigenous labor into a resource to be allocated by colonial authorities. Gold and silver extraction became priorities not because local societies needed them, but because European economies demanded them. Forests were cleared to establish plantations. Native crops were replaced with export oriented monocultures. The ecological consequences were severe, including soil exhaustion, loss of biodiversity, and increased vulnerability to famine.

Africa's experience followed the same logic under different circumstances. Long before formal colonization, African societies participated in global trade networks, developed metallurgy, agriculture, and governance systems suited to diverse environments. European colonization did not introduce complexity where none existed. It redirected existing systems toward extraction for foreign benefit. The Berlin Conference formalized this process by dividing the continent into colonies without regard for cultural or ecological boundaries. Rivers, forests, and mineral deposits were claimed as assets of distant empires.

Colonial economies were deliberately structured to prevent local self sufficiency. Railways connected mines to ports rather than towns to one another. Education systems trained intermediaries rather than innovators. Political authority was centralized to suppress dissent rather than represent communities. When independence movements succeeded, newly formed nations inherited economies designed for dependency and borders designed for conflict. The resulting instability was not a failure of post colonial governance alone. It was the predictable outcome of a system engineered for extraction rather than development.

Australia's colonization relied on an even more explicit legal fiction. The doctrine of terra nullius declared the land legally empty because Indigenous Australians did not practice agriculture or property ownership according to European norms. This erased tens of thousands of years of land stewardship that included sophisticated fire management, seasonal migration, and spiritual relationships with place. The ecological consequences of this erasure became apparent over time. Landscapes shaped by controlled burning were left unmanaged, contributing to soil degradation and increasingly severe wildfires that continue to threaten both human and non human life.

In the Pacific Islands, colonialism reached its most distilled form. Islands were valued for what could be extracted or tested rather than for the societies that lived there. Guano mining stripped ecosystems for fertilizer. Nuclear testing rendered entire atolls uninhabitable. Islanders were relocated, often permanently, in the name of military necessity and scientific advancement. The benefits accrued to distant powers, while the costs were borne locally and across generations.

Across these cases, a pattern emerges with remarkable consistency. First comes exploration framed as discovery. Then comes exploitation justified as development. Then comes extraction rationalized as progress. Finally comes inequality defended as unavoidable. This sequence is not accidental. It is the product of incentives that reward speed, ownership, and short term gain over restraint, reciprocity, and long term stewardship.

The Myths of Colonial Progress

Colonial systems did not rely solely on force. They depended on narratives that made exploitation appear reasonable, even moral. These narratives were not superficial. They shaped law, economics, and culture in ways that continue to influence how societies think about land, value, and entitlement.

One of the most powerful myths was that of emptiness. Land was declared unused if it did not conform to European models of agriculture or property. Forests were considered wasted if they were not logged. Rivers were seen as obstacles unless they powered mills or carried goods. This logic persists today in discussions of space, where asteroids are framed as meaningless until mined, and planetary surfaces are described as barren despite their scientific and cultural significance.

Another enduring myth was the civilizing mission. Colonizers claimed to bring order, religion, and progress to societies portrayed as backward or stagnant. This narrative ignored the destruction of indigenous knowledge systems and justified coercion as benevolence. In contemporary space discourse, similar language appears when technological capability is equated with moral authority, and expansion is framed as an inevitable good rather than a choice with consequences.

The promise of shared benefit formed the third pillar of colonial justification.

Infrastructure, education, and employment were cited as evidence that extraction would improve lives. In reality, these benefits were structured to support extraction rather than local resilience. Wealth flowed outward, while environmental degradation and social disruption remained. Modern arguments for space resource utilization often rely on similar assumptions, suggesting that private ownership and market mechanisms will eventually benefit all, despite historical evidence to the contrary.

Finally, the myth of trickle down prosperity presented inequality as temporary and necessary. Wealth accumulation by elites was framed as a precursor to broader development. In practice, it entrenched disparities that persist centuries later. The same logic appears in discussions of space commercialization, where concentration of ownership is defended as a stage rather than a destination.

These myths matter because they shape decisions before outcomes are visible. They determine who is heard, who is excluded, and whose losses are considered acceptable. Recognizing them is a prerequisite for avoiding their repetition.

Who Benefits from Space

The modern space economy is often presented as a shared human endeavor, yet its structure tells a different story. Access to space remains expensive, technologically complex, and politically constrained. As a result, control over space infrastructure is concentrated among a small number of nations and corporations. This concentration influences not only who profits, but who decides how space is used.

Space tourism exemplifies this imbalance. Flights beyond Earth orbit are accessible only to individuals with extraordinary wealth. These ventures are framed as inspirational milestones, yet their direct benefits accrue to a narrow demographic. Asteroid mining proposals emphasize vast potential wealth, but ownership structures suggest that profits will be captured by investors and corporations rather than distributed broadly.

Satellite infrastructure presents a more complex case. Global communication networks provide tangible benefits to millions of people, yet they also concentrate control over information and generate long term environmental risks through orbital debris. The economic gains are private, while the risks are shared. This asymmetry mirrors historical extractive industries on Earth, where profits were privatized and harms socialized.

The prospect of human settlement on Mars raises even deeper questions. Selection criteria, governance structures, and ownership models remain undefined, yet early narratives often assume that those with resources will lead. Without deliberate intervention, there is little reason to believe that space settlements will avoid reproducing existing inequalities.

History suggests that frontiers do not automatically democratize opportunity. They amplify existing power structures unless constrained by governance and ethics. Space is unlikely to be an exception.

Avoiding Cosmic Colonialism

Avoiding the repetition of colonial patterns in space is not a matter of moral purity. It is a matter of institutional design. History demonstrates that good intentions are insufficient when incentives reward extraction and speed.

International cooperation offers one path forward. The Antarctic Treaty provides a precedent for limiting exploitation in favor of scientific research and environmental protection. While imperfect and contested, it demonstrates that restraint is possible even among competing powers. Applying similar principles to space would require political will and trust that are currently fragile but not impossible.

Incorporating indigenous perspectives offers another avenue. Many cultures emphasize long term stewardship, relational ethics, and humility toward the unknown. These perspectives challenge assumptions about ownership and entitlement that dominate Western legal systems. Including them in space governance would not romanticize the past, but it would broaden the ethical framework beyond extraction.

Environmental protection must precede exploitation. Once damage occurs in space, whether through contamination or debris, it cannot be undone within human timescales. Precaution is not obstruction. It is responsibility.

Finally, speed must be recognized as a choice rather than an inevitability. Competitive narratives frame delay as weakness, yet history shows that haste benefits those already positioned to extract value. Slowing down creates space for governance, ethics, and inclusion. It allows humanity to act as a collective rather than a collection of competing interests.

In 1492, Columbus stepped onto an island believing he had reached the edge of the known world. He did not see the societies he would destroy, or perhaps he chose not to. In the coming decades, human machines and eventually human beings will set foot on other worlds. What perspectives will be missing then.

The record of colonial history is not ambiguous. The patterns are clear, the consequences documented, the justifications familiar. Every exploitation was predicted by those who were ignored. Every tragedy was foreseeable.

We already possess the textbook. The question is not whether history will judge us, but whether we will learn from what it has already taught.

13

How to Colonize Without Colonialism

Colonize is one of the most heavily burdened words in the human vocabulary. For more than five centuries it has meant the same thing in practice even when its justifications have changed. To colonize has meant to arrive with power, to define value unilaterally, to extract without consent, and to reshape land and societies according to the needs of those who arrive rather than those who are affected. Colonization has consistently presented itself as progress while producing dispossession, ecological collapse, and inequality that lasts for generations. When the word is now applied to space, it carries all of that history with it whether its speakers acknowledge it or not.

Yet space presents a genuine ethical challenge that Earth never did. No indigenous Martians will be displaced. No native civilization will be conquered. The absence of an obvious victim makes the temptation to repeat old patterns even stronger. It becomes easy to argue that because no one is visibly harmed, no harm is being done. This chapter asks whether that assumption is correct, and whether humanity is capable of settling space without importing the same mindset that has so consistently damaged every frontier it has touched.

The question is not whether humans should expand beyond Earth. Curiosity, exploration, and survival are legitimate motivations. The question is whether expansion must necessarily take the form of domination, extraction, and ownership, or whether it is possible to imagine a fundamentally different relationship between humans and the worlds they reach.

Indigenous frameworks for space

Long before rockets or satellites, many human societies developed ways of thinking about land, responsibility, and time that differed sharply from the extractive logic that dominates modern industrial culture. These frameworks did not emerge from idealism but from necessity. Communities that expected to remain in one place for generations had to learn how to live without destroying the systems that sustained them. Their survival depended on restraint, reciprocity, and long term thinking.

One of the clearest expressions of this perspective comes from the Haudenosaunee Confederacy through what is often called the Seven Generations principle. This teaching holds that every major decision should be evaluated according to its effects on the seventh generation into the future. This horizon extends far beyond individual lifespans and political terms. It forces decision makers to imagine the lived consequences of their actions on people who cannot yet speak and who will have no choice but to inherit the results.

Modern political and economic systems are structurally incapable of this kind of thinking. Corporations are rewarded for quarterly growth. Governments operate on election cycles that rarely exceed four years. Even long term infrastructure planning rarely extends beyond a few decades. The result is a culture that systematically discounts the future. Environmental damage becomes acceptable if it is delayed. Unsustainable practices persist because their costs fall outside the time frame of accountability.

Applying the Seven Generations principle to space settlement radically changes what responsible planning looks like. Designing a Mars settlement under this framework would require asking whether water extraction rates could be sustained for more than a century, not merely for the duration of an initial mission. It would require considering whether waste systems could function without contaminating the environment over generations. It would demand governance structures that could adapt over time without collapsing into authoritarian control or corporate ownership. Most importantly, it would require accepting limits rather than assuming that technological solutions will always appear in time to fix problems created by haste.

Another indigenous framework that offers profound insight for space is the Māori concept of kaitiakitanga, often translated as guardianship. Under this worldview, humans are not owners of land but caretakers with responsibilities to protect it for those who come after. Land is not inert matter but a living system with which humans are in relationship. This perspective underpinned the legal recognition of the Whanganui River as a legal entity, an acknowledgment that some parts of the world have value that cannot be reduced to ownership or utility.

Translating this concept to Mars requires a psychological shift that many find uncomfortable. It asks humans to see themselves not as conquerors of a dead world but as temporary guests with obligations. It re frames extraction as something that requires justification rather than entitlement. It places the burden of proof on those who wish to alter the environment to demonstrate that their actions will not cause lasting harm.

Closely related to guardianship is the principle of reciprocity, which appears in many indigenous cultures. Taking from the land is understood to create an obligation to give back, not as payment but as maintenance of balance. On Earth this can take the form of restoration, ceremony, or restraint. In space the question of reciprocity becomes more challenging. What does it mean to give back to a planet that is likely lifeless.

One answer is preservation. Leaving large portions of Mars untouched preserves scientific knowledge and acknowledges that not everything needs to be used to have value. Another answer is responsibility. Cleaning up waste, preventing contamination, and designing settlements that can be removed rather than permanently scarring the surface are all forms of giving back. A more abstract form of reciprocity may be cultural. Humans may bring meaning, art, and understanding to places that would otherwise never be witnessed. Whether this constitutes a true gift is debatable, but the discomfort of that debate is itself valuable because it resists the assumption that taking is cost free.

Decision making processes also matter. Many indigenous societies rely on council based governance that emphasizes consensus and deliberation over speed. Such systems are slow by design, not because they are inefficient but because they prioritize preventing irreversible mistakes. In contrast, corporate and military driven space planning emphasizes rapid execution and competitive advantage. Translating council based decision making to space governance would mean accepting that some actions should take longer to approve because their consequences are permanent.

The Antarctic model

Humanity has already demonstrated that it is capable of restraint beyond Earth under the right conditions. The Antarctic Treaty stands as one of the most remarkable examples of international cooperation in modern history. Signed during the height of the Cold War, it set aside an entire continent for peaceful scientific use, prohibited military activity, and froze territorial claims. Later agreements strengthened environmental protections, banned mining, and established strict waste management requirements.

The success of the Antarctic Treaty lies not in perfection but in structure.

Clear rules were established early, before large scale exploitation began. Scientific cooperation was prioritized over competition. Environmental protection was treated as a core objective rather than an obstacle. Decisions were made collectively and by consensus, which prevented unilateral exploitation.

There are important limitations to this model. Antarctica has no permanent human population and historically lacked immediately exploitable resources, which reduced incentives to cheat. Enforcement relies heavily on transparency and peer pressure rather than force. Tourism and climate change have introduced challenges that the original treaty did not anticipate. Despite these flaws, the treaty has largely succeeded in preventing the kind of extractive free for all that has characterized other frontiers.

Applying this model to space suggests that early, binding agreements are essential. Waiting until exploitation is profitable almost guarantees failure. A space environmental treaty modeled on Antarctica but adapted for commercial activity could mandate environmental impact assessments, designate protected zones, and require shared scientific data. Crucially, it would need enforcement mechanisms with real consequences for violations, including loss of operating licenses and economic penalties.

Alternative governance models

Several governance frameworks could support ethical space settlement if political will exists. One option is a form of international trusteeship under the United Nations, where celestial bodies are held in trust for all humanity. Such a body could authorize activities, designate protected areas, and manage benefit sharing. While critics argue that the UN is slow and politicized, its legitimacy and exclusivity offer advantages over unilateral national control.

Another option is a specialized international authority modeled on agencies that regulate nuclear materials or aviation safety. A Space Environment Authority could set standards for debris mitigation, contamination prevention, and sustainable resource use. It could monitor compliance using space tracking systems and coordinate cleanup efforts when damage occurs. While enforcement would still depend on member states, standardized licensing could exert significant pressure on commercial actors.

A more experimental model involves rotating stewardship. Under this system, no nation or corporation would hold permanent control over a celestial body. Stewardship would rotate among international groups, with each steward evaluated on environmental and ethical performance. This approach emphasizes responsibility over ownership and reduces the incentive to extract as quickly as possible.

Benefit sharing is essential under any model. Space resources cannot ethically become the exclusive property of those with the technology to reach them. Mechanisms such as resource taxes, shared technology access, and open scientific data can help ensure that space activity benefits humanity as a whole rather than a narrow elite. Precedents for such systems exist in international maritime law and global scientific collaborations.

The cultural shift required

Governance structures alone are insufficient if the underlying mindset remains unchanged. Language shapes perception, and much of the language surrounding space still echoes conquest. Phrases such as conquering space or claiming territory reinforce the assumption that domination is natural and desirable. Shifting toward language of stewardship, settlement, and partnership signals a deeper change in values.

Education plays a central role in this shift. Current space education focuses heavily on engineering and technology while treating ethics as an afterthought. Integrating space ethics into curricula at all levels would help future engineers, scientists, and policymakers internalize responsibility alongside capability. Just as medical training includes ethics because lives are at stake, space training must include ethics because entire environments are at stake.

Art and storytelling are equally important. The futures that societies imagine influence the futures they build. Science fiction that explores cooperative, sustainable space societies can make alternatives to exploitation feel plausible. Indigenous futurism and non Western perspectives can broaden the narrative beyond familiar tropes of conquest.

At a deeper level, space challenges humanity to reconsider its place in the universe. A mindset that views the cosmos as raw material inevitably leads to exploitation. A mindset that sees humanity as part of a larger unfolding system encourages humility and care. Many philosophical and spiritual traditions emphasize interconnection and restraint. Bringing these perspectives into space discourse does not require abandoning science or progress. It requires redefining what progress means.

Ending

Humanity has spent centuries perfecting the mechanics of expansion while neglecting its ethics. The result is a planet under strain and a history filled with avoidable harm. Space offers a rare opportunity to choose differently, not because the stakes are lower but because the record is clearer.

Colonizing without colonialism is not a contradiction if colonization is redefined. Settlement does not have to mean domination. Expansion does not have to mean extraction without limits. Growth does not have to mean inequality. The frameworks exist. Indigenous wisdom, international cooperation, environmental science, and ethical governance all provide tools for a different approach.

What remains uncertain is whether humanity has the will to use them. Speed will always be tempting. Profit will always exert pressure. Competition will always encourage shortcuts. Resisting these forces requires conscious choice rather than default behavior.

The universe does not need humans to arrive. If humans choose to go, the moral burden rests entirely on them. The question is no longer whether humanity can reach the stars. The question is whether it can do so without repeating the same mistakes on a larger stage.

That choice is not theoretical. It is being made now, in policies, investments, and mission designs. History has already shown what happens when expansion proceeds without restraint. Space offers no excuse for ignorance, only a test of whether humanity has learned enough to deserve a future among the stars.

14

The Green Space Station

The International Space Station recycles roughly ninety three percent of its water, a fact that initially provokes discomfort before it inspires awe. Urine becomes drinking water through careful filtration. Sweat condenses into humidity that is captured and purified. Even the moisture from breath becomes tomorrow's hydration. What sounds unpleasant is in fact a triumph of engineering and necessity working together. Yet this remarkable achievement still falls short of what deep space survival requires, because the remaining fraction must be delivered from Earth at staggering cost. For missions beyond low Earth orbit, resupply is not merely expensive but impossible. When there is no cargo ship coming, recycling must approach perfection or life simply ends.

This reality forces a deeper question than engineering alone can answer. If survival in space depends on total recycling, then perhaps the model we have been using is incomplete. Machines can filter and scrub, but machines wear out. Filters clog. Pumps fail. Replacement parts weigh kilograms that must be lifted out of Earth's gravity well. Nature, by contrast, has been operating closed systems for billions of years without spare parts or external resupply. Forests, reefs, and soils recycle matter endlessly through relationships rather than components. If humans are going to live in space for more than short visits, the design philosophy must change. Habitats must begin to function less like factories and more like ecosystems.

The green space station is not a metaphor. It is a requirement imposed by physics, biology, and economics at the same time. Sustainability in space is not a moral preference but a condition of survival. The question is not whether we want to live this way, but whether we can afford not to.

Biomimicry for space

Earth is the only example of a fully functioning closed life support system that humanity has ever known. For more than four billion years, it has recycled the same atoms through oceans, rocks, air, and living organisms. Aside from a steady input of energy from the Sun, almost nothing enters and nothing leaves. Water is not lost but transformed. Carbon is not consumed but rearranged. Nitrogen circulates endlessly through invisible microbial labor. The planet does not operate through centralized control or rigid components. It operates through networks of relationships that adapt and repair themselves.

The water cycle illustrates this elegance. Liquid water evaporates into vapor, leaving impurities behind. That vapor condenses into clouds and returns as precipitation, purified by the process itself. Rivers carry water back to the sea, where the cycle begins again. No pump is required, no filter replaced. Gravity and phase change do the work. In space habitats, engineers already attempt to mimic this logic through distillation and condensation, but they often stop short of the broader lesson. The power of the cycle lies not only in the mechanism but in its integration with everything else.

The carbon cycle offers another essential blueprint. Plants take in carbon dioxide and use energy from light to build sugars and release oxygen. Animals consume oxygen and organic matter, returning carbon dioxide and waste. Microbes and fungi decompose that waste, freeing nutrients for plants once more. At the scale of an ecosystem, there is no waste, only material waiting to be reused. When humans disrupt this balance on Earth, carbon accumulates in the atmosphere and climate shifts. In space, imbalance would be fatal far more quickly.

Nitrogen cycling demonstrates the importance of organisms that humans rarely notice. Most life cannot use atmospheric nitrogen directly. Specialized bacteria perform the work of fixing it into forms that plants can absorb. Without these microbes, proteins could not be built and complex life would cease. Any closed space habitat that ignores microbial communities is incomplete by definition. Life support does not begin with humans or plants but with bacteria.

What unites these cycles is not efficiency in the mechanical sense but resilience. Earth does not depend on a single process or species. Multiple organisms perform overlapping roles. When one pathway falters, others compensate. Engineers call this redundancy. Ecologists call it diversity. In space, redundancy cannot come from spare parts alone. It must come from living systems capable of adaptation.

Nature offers countless examples of design principles that can inform space habitats. Termite mounds in Africa maintain stable internal temperatures despite extreme heat outside. They achieve this through geometry rather than electricity. Carefully placed vents create convection currents that move hot air out and draw cool air in. The mound itself becomes a climate control system. For space habitats where energy is precious and failure dangerous, passive thermal regulation is not a luxury but a safeguard.

Lotus leaves provide another lesson. Their surfaces are structured at microscopic scales that cause water to bead and roll off, carrying dirt with it. This self cleaning property has already inspired coatings on Earth. In space, where dust accumulation can cripple solar panels and windows, such surfaces reduce maintenance and risk. Each repair avoided is an extravehicular activity not required, and each avoided excursion reduces danger.

Coral reefs demonstrate the power of symbiosis. Coral animals host algae within their tissues. The algae provide energy through photosynthesis, while the coral provides shelter and nutrients. Neither survives alone for long. Together they build structures that support thousands of other species. A space habitat must be designed around similar mutual dependencies. Humans produce waste gases and organic matter that plants and microbes require. Plants produce oxygen and food that humans need. Treating these flows as problems to eliminate rather than relationships to cultivate is a fundamental mistake.

Even death is not waste in nature. When a whale dies and sinks to the ocean floor, its body becomes the foundation of a community that can last decades. Soft tissue feeds scavengers. Bones support bacteria that draw energy from chemical reactions. Nothing is discarded. In space, every scrap of organic matter must be treated with the same respect. Food waste, human waste, and dead plants are not liabilities. They are resources temporarily out of place.

The shift required is conceptual. Traditional life support systems treat biology as something to be controlled or supplemented. A bioregenerative system treats biology as the core, with machines as support. Mechanical systems are precise but brittle. Biological systems are messy but resilient. In the isolation of deep space, resilience matters more than precision.

Sustainable technologies

Energy underpins every closed system. In orbit around Earth, solar power provides reliable electricity. Without an atmosphere, sunlight is stronger and uninterrupted by weather. The International Space Station generates over one hundred kilowatts through its arrays, enough to sustain life and research. For habitats near Earth or Mars, solar remains viable, though dust accumulation and reduced intensity at greater distances require adaptation.

Beyond Mars, the physics of distance impose limits. Sunlight weakens with the square of distance, making solar impractical in the outer solar system. There, nuclear power becomes essential. Radioisotope generators have already powered missions for decades, converting heat from radioactive decay into electricity. Their output is small but reliable. Compact fission reactors promise greater power for habitats, operating continuously for years. Though controversial, their safety record in space has been strong, and without them sustained presence far from the Sun would be impossible.

Energy alone does not sustain life. Materials must be sourced locally whenever possible. In situ resource utilization transforms hostile environments into livable ones by using what is already present. On Mars, radar surveys confirm vast deposits of subsurface ice. That water can be melted for drinking, split into oxygen and hydrogen, and combined with atmospheric carbon dioxide to create fuel. Experiments have already demonstrated these reactions at small scale. Scaling them up is an engineering challenge, not a scientific mystery.

Martian soil can be turned into building material through heating or binding processes, creating bricks or printed structures that shield against radiation. Metals can be extracted from oxides and reused. Each kilogram produced locally is a kilogram not launched from Earth. On the Moon, permanently shadowed craters contain ice that can support fuel depots and habitats. Asteroids rich in water and metals could supply entire orbital economies. The logic is simple. Permanent settlement requires local sourcing. Anything else is temporary by definition.

At the heart of the green space station lies the closed ecological life support system. Plants are not decorative. They are essential machinery that grows rather than degrades. Through photosynthesis, they produce oxygen and food while recycling carbon dioxide and water. Crops selected for space must balance nutrition, yield, and resilience. Leafy greens grow quickly and provide vitamins. Tubers offer calories. Grains supply carbohydrates. Algae provide protein and micro nutrients with extraordinary efficiency.

Growing plants in space requires abandoning soil in favor of controlled systems. Hydroponics delivers nutrients directly through water. Aeroponics suspends roots in air and mists them, reducing water use even further. Aquaponics integrates fish, whose waste feeds plants while plants clean the water. Each method reflects the same principle. Outputs become inputs elsewhere.

Microbes are the quiet backbone of these systems. Decomposers break down waste into forms plants can use. Nitrogen fixing bacteria make essential nutrients available. Beneficial microbes support human health in environments where microbial diversity would otherwise decline. Without a thriving microbiome, a habitat will fail regardless of how advanced its machinery is.

Animals are optional but valuable. Fish provide complete protein and enhance nutrient cycling. Insects can convert waste into edible biomass with remarkable efficiency, though cultural acceptance varies. Larger animals consume too many resources to justify their presence nutritionally, but their psychological benefits should not be dismissed lightly. Human well being is part of sustainability.

Balancing such a system is difficult. Too many plants can raise oxygen to dangerous levels. Too few lead to suffocation. Waste must be processed at the right rate. Diversity is key. Mono cultures collapse under stress. Redundancy provides insurance. Experiments on Earth have demonstrated both failure and success. Early attempts struggled with unexpected chemical interactions and social strain. More recent trials have achieved near total closure for extended periods. These results show that the challenge is not possibility but patience and design.

Zero waste space living

On Earth, waste disappears from view, giving the illusion that it no longer exists. In space, that illusion vanishes. Every molecule remains inside the system. Thermodynamics guarantees it. Economics reinforces it. Launching mass costs thousands of dollars per kilogram. Replacing nutrients lost through waste becomes prohibitively expensive. In a closed environment, waste is simply matter in the wrong place.

The International Space Station already treats water as too valuable to discard. Its recovery system captures moisture from nearly every source and purifies it through multiple stages. Oxygen is generated by splitting water and reclaimed through chemical reactions. Yet solid waste and packaging still accumulate and must be removed. For deep space habitats, this approach is insufficient.

Future systems must process all waste internally. Human waste can be composted biologically, gasified chemically, or digested by microbes to produce fertilizer and fuel. Food scraps can feed worms or insect larvae, which in turn feed fish or humans. Plastics can be broken down and reprinted. Metals can be melted and reused. The goal is not perfection in any single process but closure across the whole.

A day in such a habitat would feel unremarkable to its inhabitants once routines settle. Water flows in loops unseen. Food grows quietly. Waste disappears into systems that return nutrients to the soil substitute. Nothing leaves except heat radiated into space. The absence of trash is not austerity. It is efficiency.

Ending

The green space station represents a convergence of necessity and insight. Nature has already solved the problem of living in closed systems. Humans must learn to listen rather than impose. Biomimicry provides designs refined over billions of years. Bioregenerative systems turn vulnerability into strength. Sustainable technologies align with physical reality rather than resisting it.

The implications reach beyond space. Systems developed to survive on Mars mirror those needed to heal Earth. Water recycling, closed loop agriculture, waste elimination, and renewable energy are not luxuries in space. They are requirements. Their adoption on Earth would be wise rather than merely ethical.

Humanity often imagines space as an escape from environmental limits. In truth, space enforces those limits with absolute clarity. There is no excess to exploit, no away to discard into, no safety net of resupply. Living among the stars requires maturity, restraint, and respect for cycles that do not forgive neglect.

A future habitat on Mars filled with plants, microbes, animals, and people living in balance is not fantasy. It is a logical outcome of aligning design with reality. Whether that outcome is achieved depends on choices made now. Sustainability cannot be added later. It must be built from the beginning or not at all.

The green space station is not simply a structure. It is a test of whether humanity can learn from nature rather than overpower it. Passing that test may determine not only how far humans travel, but how long they endure wherever they go.

15

Terra forming vs. Preservation

Imagine you discover the Mona Lisa in a dusty attic. The paint is cracked, the colors muted by centuries of time, but it is unmistakably authentic. Five hundred years of history stare back at you from the canvas. You have a choice. You can preserve it exactly as it is, study it, protect it, and pass it on intact to the future. Or you can paint over it, adding brighter colors, correcting what you see as flaws, reshaping it into something more pleasing to your own tastes.

Now imagine Mars. A planet four and a half billion years old. Ancient river valleys carved when Earth was still young. Polar ice caps layered with climate history. The only other world in our solar system where humans might one day walk under an open sky. Again, the choice presents itself. Do we preserve it, study it, and touch it lightly, or do we transform it completely, warming it, thickening its atmosphere, and reshaping it into a second Earth.

One choice protects history. The other destroys history in order to create a future. There is no neutral option.

This is not a distant thought experiment. Space agencies and private companies are planning Mars missions on timelines measured in decades, not centuries. Engineering papers explore ways to warm the planet. Conferences debate how to thicken the atmosphere. Some scientists argue that we should begin terraforming as soon as possible, releasing greenhouse gases and triggering a slow planetary transformation. Others argue that doing so would be the greatest act of environmental vandalism humanity has ever committed. The window for deciding is narrowing, and most people do not even realize that the decision exists.

The Terra forming dream

Terra forming is a word that carries both scientific precision and mythic weight. The term comes from the Latin word for Earth and the idea of shaping. To Terra form a planet is to reshape it until it resembles Earth, with a breathable atmosphere, liquid water on the surface, and temperatures compatible with human life. It is the ultimate expression of planetary engineering, an ambition that moves beyond cities and ecosystems to the scale of worlds.

Mars is the obvious candidate for this dream. It already possesses many of the ingredients that make terraforming conceivable, if not easy. Its day length is similar to Earth's. Its axial tilt produces seasons. It has ice locked away at the poles and beneath the surface. Long ago, rivers flowed across its surface and lakes filled its basins. In many ways, Mars feels like a planet that almost made it.

Today, however, Mars is profoundly hostile to unprotected human life. Its atmosphere is so thin that it is essentially a vacuum by Earth standards. Oxygen makes up a tiny fraction of that already scarce air, far too little to breathe. The low pressure means that exposed bodily fluids would boil at normal body temperature. Radiation from the Sun and distant cosmic sources reaches the surface largely unfiltered because Mars lacks a global magnetic field. Temperatures swing wildly, from pleasant warmth at midday near the equator to lethal cold at night and near the poles. The soil itself contains chemicals that are toxic to humans.

Because of these conditions, every current plan for Mars exploration relies on sealed habitats, pressure suits, and constant life support. Humans would not live on Mars so much as survive inside carefully maintained bubbles. A single failure could be fatal. Terraforming promises something radically different. It promises a Mars where humans could step outside without a suit, breathe the air, feel rain, and plant crops directly in the soil. It promises not survival, but life.

The vision is intoxicating. It extends far beyond the lifetimes of those who first propose it. Advocates ask us to imagine future generations born on Mars, who have never seen Earth except in pictures. They would run through fields under a Martian sky, swim in Martian seas, and call Mars home in a way that no astronaut ever could. Humanity would no longer be confined to a single fragile world. We would be a multi planetary species in the deepest sense.

Turning that vision into reality would require changes on a planetary scale. The first step would be warming Mars. On average, the planet is far too cold for liquid water to exist for long on its surface. To change that, Mars would need to be heated by tens of degrees. Proposals range from placing enormous reflective mirrors in orbit to concentrate sunlight onto the polar ice caps, to building factories that release powerful greenhouse gases into the atmosphere. Some have even suggested redirecting icy asteroids to collide with Mars, delivering both heat and volatiles. A few proposals have gone so far as to suggest detonating nuclear devices at the poles to release trapped carbon dioxide, though this idea raises obvious political and environmental concerns.

As Mars warms, frozen carbon dioxide would sublimate into gas, thickening the atmosphere and amplifying the greenhouse effect. In theory, this could create a positive feedback loop, where warming releases more gas, which causes further warming. Over decades or centuries, the atmospheric pressure could rise from its current near vacuum to a level where liquid water could persist on the surface.

Once temperatures and pressures crossed critical thresholds, ice would begin to melt. Rivers could flow again. Lakes could fill basins that have been dry for billions of years. A hydrological cycle might emerge, with evaporation and precipitation reshaping the landscape. Mars would begin to resemble its ancient self, at least superficially.

Life would come next, introduced intentionally. The first organisms would likely be microbes adapted to extreme conditions. Photosynthetic bacteria could begin producing oxygen, while nitrogen fixing microbes could make the soil more fertile. Over time, simple plants might follow, gradually stabilizing the environment and accelerating atmospheric change. Given enough time, the oxygen content of the atmosphere could rise to levels suitable for humans and animals.

The timescales involved are immense. Even the most optimistic estimates suggest centuries to millennia before Mars could support unprotected human life across much of its surface. Terraforming is not a project for individuals or even nations. It is a commitment that binds generations yet unborn.

Those who advocate for terraforming do so for reasons that are both practical and philosophical. Some argue that humanity needs a second habitable planet as insurance against extinction. Earth faces risks from climate change, asteroid impacts, pandemics, and self inflicted disasters. A second world could ensure the long term survival of our species. Others see terraforming as the ultimate scientific and engineering challenge, one that could yield knowledge applicable to managing Earth's climate and ecosystems. Still others frame it as a moral imperative, arguing that life has a tendency to spread and that transforming a barren world into a living one is inherently good.

Underlying all these arguments is a belief that Mars is, in essence, empty. If no life exists there now, then terraforming harms nothing. From this perspective, Mars is a canvas waiting for life, and humans have the tools to paint.

The ethical problems

Against this powerful vision stand equally compelling objections. To many scientists, Mars is not an empty canvas but an irreplaceable archive. Unlike Earth, Mars has no active plate tectonics to recycle its crust. Its surface preserves a record of planetary history stretching back billions of years. Ancient river deltas, sedimentary layers, and mineral deposits hold clues to how planets form, how climates change, and how life might emerge or fail to do so.

Once terraforming begins, that record would be irreversibly altered. Flowing water would erode and bury ancient features. Introduced life would contaminate chemical signatures that scientists rely on to detect past biology. Human activity would disturb sites that have remained unchanged for eons. The pristine Mars that holds answers to fundamental questions about our own origins would be lost forever.

This loss is not hypothetical. Humanity has already destroyed much of Earth's early geological record, making it difficult to reconstruct the conditions under which life first arose here. Mars offers a rare second chance to study a planet that diverged from Earth early in its history. To erase that record before fully understanding it would be a profound act of scientific negligence.

The life question deepens the ethical stakes even further. Despite decades of exploration, we cannot say with certainty that Mars is lifeless. Our searches have been limited to a tiny fraction of the surface, and almost none of the subsurface has been explored. On Earth, life thrives in environments once thought impossible, from deep underground to within solid rock. It is entirely plausible that microbial life could persist on Mars in sheltered niches, surviving in ice or briny water far below the surface.

If such life exists, terraforming would almost certainly destroy it. Changing temperature, pressure, and chemistry would alter habitats that Martian organisms have adapted to over billions of years. Introducing Earth microbes would create competition that native life might not survive. Even if no complex organisms exist, extinguishing a second genesis of life would be a moral catastrophe of unprecedented scale.

The precautionary principle suggests that when the stakes are this high, uncertainty should counsel restraint. The benefit of terraforming is speculative and distant, while the harm of destroying alien life would be immediate and irreversible. Is it acceptable to risk annihilating the only other life we might ever find in order to accelerate human expansion?

Ir reversibility is perhaps the most troubling aspect of terraforming. Once begun, planetary scale changes cannot be undone on any meaningful human timescale. If future generations decide that the choice was wrong, they will have no way to restore Mars to its original state. Decisions made by a small number of people today would bind countless generations to come.

This raises a profound governance problem. There is no international body with clear authority over terraforming. Existing space treaties were written in an era when such capabilities were science fiction. If a single nation or corporation gains the technological ability to begin warming Mars, nothing currently prevents it from acting unilaterally. The idea that one individual or organization could reshape an entire planet that belongs, in principle, to all humanity is deeply unsettling.

History offers uncomfortable parallels. Colonial powers once transformed lands without the consent of those who lived there, confident in their own righteousness. The consequences of those decisions are still felt today. Terraforming Mars without broad global consensus risks repeating this pattern on a cosmic scale.

An alternative path: paraterraforming

Between total transformation and strict preservation lies a third option. Paraterraforming proposes that humans create Earth like environments in limited, enclosed spaces on Mars, rather than altering the entire planet. Instead of warming Mars itself, we would build habitats that provide breathable air, comfortable temperatures, and familiar ecosystems inside domes, underground complexes, or sealed valleys.

This approach has deep roots in human history. People have always created microclimates to survive in hostile environments. Igloos in the Arctic, underground cities in deserts, submarines beneath the ocean, and space stations in orbit all follow the same principle. Humans adapt their surroundings locally rather than forcing the entire environment to change.

On Mars, paraterraforming could take many forms. Large transparent domes could enclose cities, allowing sunlight in while maintaining Earth like conditions inside. Deep canyons could be partially sealed to create vast protected spaces shielded from radiation. Underground habitats could use the planet's crust as insulation, providing stability and safety.

Such habitats would allow people to live, work, and even farm without pressure suits, while leaving the vast majority of Mars untouched. Scientific exploration could continue outside the habitats, preserving the planet's natural state. If future discoveries revealed the presence of life, it could be protected. If, after centuries of study, humanity decided that full terraforming was justified, that option would remain open.

Paraterraforming has practical advantages as well. It is achievable on much shorter timescales, using technologies that already exist or are under development. It is reversible in principle, allowing future generations to change course. Failures would be localized rather than planetary. Ethical responsibility becomes clearer when the scale of intervention is limited.

The drawbacks are real. Life inside domes would always be artificial to some degree. Expansion would be constrained by construction rather than planetary abundance. The psychological impact of living under a roof on an alien world is not trivial. Yet compared to the permanence and uncertainty of terraforming, these challenges may be more acceptable.

The deeper question

At its core, the debate between terraforming and preservation is not only about Mars. It is about how humans see themselves in relation to nature. Terraforming embodies a mindset of control. When the environment does not suit us, we change it. This approach has powered remarkable achievements, from agriculture to cities to modern technology. It has also driven environmental degradation and climate change on Earth.

Preservation and paraterraforming reflect a mindset of acceptance and adaptation. Instead of reshaping the world to fit human needs, humans reshape themselves and their tools to fit the world. This perspective aligns more closely with many indigenous worldviews, which emphasize relationship and respect rather than domination.

Neither mindset is inherently wrong. Humanity would not exist without the ability to modify environments. Yet the scale of terraforming magnifies the consequences of error beyond anything we have faced before. Earth's environmental crisis demonstrates how difficult it is to manage complex systems, even with the best intentions. Mars offers no safety net.

Some frame terraforming as playing god, an act of hubris that exceeds our moral authority. Others see it as fulfilling human potential, an expression of creativity and responsibility. Religious and philosophical traditions diverge sharply on this point, and no argument conclusively resolves the tension.

Asking what Mars wants may sound absurd, since planets are not conscious. Yet the question gestures toward a deeper issue of intrinsic value. Does Mars matter only insofar as it serves human purposes, or does its long history confer a value that demands respect regardless of utility. There is no scientific answer to this question. It is a matter of ethics and worldview.

Ending

After exploring the arguments, it becomes clear that there is no simple resolution. Terraforming offers the promise of a second living world and the long term survival of humanity. Preservation offers knowledge, humility, and protection against irreversible harm. Paraterraforming offers a compromise, trading total freedom for caution and reversibility.

What makes the choice so difficult is not ignorance but uncertainty. We do not know whether Mars harbors life. We do not know whether terraforming would work as planned. We do not know what future generations would value more, a transformed Mars or an untouched one. We are making decisions with incomplete information, as humans always do.

Three principles emerge as guides rather than answers. First, study must come before transformation. Mars deserves decades, perhaps centuries, of thorough exploration before any irreversible action is taken. Second, limited and reversible approaches should precede planetary scale commitments.

Living on Mars does not require remaking it. Third, no single nation, company, or individual should decide the fate of an entire planet. The decision must involve humanity as a whole, including voices that have historically been excluded from such choices.

Whether these principles will be followed is uncertain. The technological capability to begin altering Mars may arrive before the political or ethical frameworks to govern that power. If that happens, history suggests that action will precede reflection.

Perhaps the most important question, then, is not whether we should terraform Mars, but whether we are wise enough to do so. Wisdom is not the same as intelligence or capability. It is the ability to recognize limits, to value patience, and to act with humility in the face of complexity.

Earth's environmental crisis suggests that humanity is still learning that lesson. Mars will test whether we have learned enough.

16

Two Paths Forward

We stand at a fork in the cosmic road, and this fork is not a metaphor meant to sound poetic after the fact. It is a real divergence that is forming right now, in the decade you are living through, shaped by decisions that rarely make headlines and almost never involve public consent. The path splits here in the 2020s and 2030s, not in some distant era when humanity has more wisdom or better technology, but in the present moment while rockets are being ordered, contracts signed, and missions approved. It is happening in boardrooms where cost cutting is framed as efficiency, in government committees where regulation is called a burden, and in technical conferences where environmental concerns are politely acknowledged and quietly sidelined.

Both paths lead outward into space. Both promise expansion, progress, and survival beyond Earth. Both sound reasonable when described in isolation. Yet they lead to radically different futures. One path leads to a sustainable and enduring human presence in the cosmos, one built on restraint, cooperation, and respect for environments older than humanity itself. The other path leads to catastrophe, confinement, and a gradual collapse of space exploration under the weight of our own negligence. The most unsettling part is not that the choice is difficult, but that most people do not know the choice exists at all.

The timelines are already diverging. To understand what is at stake, we must walk them forward and see where they lead, not in abstract terms but through concrete events, lived experiences, and the cumulative consequences of thousands of small decisions.

Timeline A, the extraction future

2028: the first commercial asteroid mining mission

The launch takes place on a humid morning at Cape Canaveral, surrounded by a crowd that includes journalists, investors, and politicians eager to attach their names to a historic moment. Astro Mine Industries, a venture backed startup with two billion dollars in funding, sends its first robotic spacecraft toward Bennu, a near Earth asteroid roughly five hundred meters across. The stated mission is prospecting, with a focus on platinum group metals and water that can be turned into rocket fuel. News outlets describe it as the beginning of a new gold rush, a moment when space finally becomes economically useful rather than merely symbolic.

In the months that follow, the problems emerge quietly. The spacecraft design relies on explosive separation bolts, a mature but messy technology chosen because it is cheap, lightweight, and well understood. Each detonation produces fragments, some large enough to track and many too small to see. There is no deorbit plan for spent stages or discarded components, because the regulatory framework does not require one. Debris begins accumulating in the Earth Moon corridor, an increasingly important region for future missions.

On board the International Space Station, Commander Maya Chen receives a debris alert during what should have been an uneventful orbit. The crew retreats into their return capsule as procedure demands, waiting for a fragment that ultimately passes two hundred meters away. The margin sounds large until one remembers the speed involved, where a marble sized object carries the energy of a grenade. Chen files a formal complaint through commercial oversight channels, arguing that AstroMine is treating orbital space as a disposable worksite. The complaint is acknowledged and then effectively ignored, lost in a jurisdictional fog where no agency has clear authority over private activity beyond Earth orbit.

The precedent is subtle but decisive. A private company has demonstrated that it can pollute space without meaningful consequence, and competitors take note. Speed becomes the dominant value. First mover advantage outweighs long term responsibility. The logic that once governed social media platforms, move fast and fix later, migrates seamlessly into space.

2033: Mars base established and the water conflicts begin

Humanity's first permanent settlement on Mars is established in Jezero Crater, a location chosen for its geological interest and proximity to subsurface ice. Forty seven people arrive under banners of international cooperation, yet within months the cooperative language gives way to competitive reality. Missions from multiple nations land in the same general region, all drilling for the same limited water deposits. There is no binding coordination protocol, no enforceable agreement on shared use, and no neutral authority to resolve disputes.

Dr. James Okoye, a Nigerian geologist working with an international science team, notices unusual chemical patterns in ice cores extracted during early drilling. The signatures are ambiguous but intriguing, suggesting processes that could be biological rather than purely chemical. He urges caution and requests time for careful analysis, pointing out that contamination would permanently compromise the data. Mission control declines, citing the immediate need for water to sustain the settlement and produce fuel. Schedules matter, investors are watching, and delays are framed as existential risks to the mission.

Within weeks, contamination protocols are relaxed. Equipment that has never been fully sterilized breaches subsurface ice layers that have been isolated for millions or perhaps billions of years. Earth microbes enter environments that may once have supported Martian life, or may still do so in some dormant form. Later analyses reveal ambiguous bio signatures that can never be definitively classified as native or terrestrial. Scientific value that could have answered fundamental questions about life in the universe is lost in the rush to secure resources.

Territorial disputes soon follow. Safety zones become de facto claims. Flags are planted. Diplomatic protests are issued and quietly ignored. The Outer Space Treaty prohibits sovereignty claims, but it offers no enforcement mechanism, and power follows presence. Mars begins to resemble Earth's colonial past, where whoever arrives first defines the rules.

2039: the Kessler cascade begins

At the Space Surveillance Center in Colorado, Lieutenant Colonel Sarah Martinez watches a collision unfold in real time. A Starlink satellite collides with a defunct Chinese spacecraft at nearly fifteen kilometers per second. The impact produces thousands of tracked fragments and millions of smaller ones, spreading across multiple orbital planes. Within days, secondary collisions occur as debris strikes other satellites. Each impact multiplies the problem.

Avoidance maneuvers increase sharply. Insurance premiums soar. Some operators deorbit functioning satellites to reduce risk, but the effort comes too late. The density of debris crosses a critical threshold in several orbital bands. Low Earth orbit becomes a high speed minefield.

Within a year, the International Space Station is evacuated under emergency conditions. Weather satellites fail in clusters, degrading storm prediction and climate monitoring. Navigation systems lose precision. Communications flicker and fail. Economists estimate losses in the trillions as satellite dependent industries scramble to adapt. The debris will not decay for decades, and cleanup is deemed too expensive and politically unpalatable. Humanity has wrapped its planet in a shell of shrapnel.

2045: the lunar south pole is stripped

Industrial mining begins at Shackleton Crater, one of the Moon's oldest and coldest regions. Permanently shadowed craters that have preserved water ice for billions of years are excavated at scale. The operation processes ten thousand tons of regolith per day, extracting water for rocket fuel destined for Mars missions.

Dr. Kenji Tanaka, a lunar geologist who has studied the Moon for decades, arrives for an inspection and finds a landscape transformed beyond recognition. Craters are leveled, ice deposits destroyed, and historical Apollo sites threatened by expansion. The climate record encoded in those ice layers, a unique archive of solar system history, is gone before it can be studied. The scars become visible from Earth through amateur telescopes, prompting public outrage that arrives far too late to matter.

2052: Europa's ocean is contaminated

A long anticipated mission drills through Europa's icy shell to reach its subsurface ocean. Budget pressures and political impatience shorten preparation timelines. Contamination safeguards are weakened to save cost and mass. Dr. Amara Okafor, an astrobiologist and daughter of James Okoye, raises concerns that are acknowledged and dismissed.

Initial readings from the deployed submarine detect organic molecules, and excitement spreads rapidly through mission control. Weeks later, genetic analysis reveals unmistakable Earth bacterial signatures. The realization is devastating. Humanity cannot know whether Europa's ocean ever hosted native life or still does. If it did, it may now be extinct or irrevocably altered. The greatest scientific question of all time is rendered unanswerable through haste.

2061: terraforming begins without consent

China announces the deployment of orbital mirrors around Mars to begin warming the polar regions. International protests follow, but no law prevents the action. Carbon dioxide ice begins to sublimate, thickening the atmosphere slightly and initiating a process that cannot be reversed on human timescales.

Dr. Sarah Okafor, now elderly and long retired, watches the announcement from a nursing home. She has spent her life arguing for caution and preservation. Her work is undone by a single unilateral decision. Mars is no longer pristine, and whatever it once was, it will never be again.

2075: collapse and retreat

The Mars colony, once celebrated as humanity's boldest achievement, is abandoned. Water reserves are depleted. Supply chains falter as launches through debris filled orbit become increasingly dangerous. Life support failures cascade. Forty seven people die during the final evacuation.

By 2075, low Earth orbit is largely unusable, the Moon is scarred, Mars is altered but uninhabitable, and Europa's secret is lost forever. Public support for space exploration collapses. Space is remembered not as a frontier of possibility, but as a cautionary tale of unchecked ambition.

The pattern is complete. Exploration gives way to exploitation, exploitation accelerates extraction, extraction triggers environmental collapse, and collapse leaves abandonment behind. Humanity exports its worst habits to the cosmos and discovers that space, like Earth, cannot absorb them without consequence.

Timeline B, the stewardship future

2026: the space environment treaty

At the United Nations, representatives from one hundred forty seven nations sign the International Space Environment Treaty. Modeled on Antarctic governance but far stronger in enforcement, it establishes protected zones, mandatory debris mitigation, strict contamination protocols, and benefit sharing mechanisms. Ambassador Amara Diop of Senegal plays a key role in ensuring that developing nations have a voice, arguing that space must benefit all humanity rather than a handful of wealthy actors.

The treaty marks a historic shift. Environmental protection is established before large scale exploitation begins, reversing the pattern that defined Earth's industrial history.

2030: Mars is mapped before it is touched

Robotic missions survey Mars extensively before any humans land. Potential bio signature sites are identified and protected. Ninety percent of the planet is designated off limits to settlement and extraction. When Dr. James Okoye detects unusual chemistry in subsurface ice, the area is sealed as a scientific preserve. The decision costs time and money, but it preserves the possibility of discovery rather than foreclosing it.

2035: a sustainable lunar presence

A small research station operates near Shackleton Crater, harvesting water carefully while leaving most ice untouched. Life support systems recycle nearly everything. Engineer Isabella Martinez sees her bio regenerative designs outperform purely mechanical systems, demonstrating that sustainability is not a constraint but an advantage. The Moon remains largely unchanged, studied rather than consumed.

2048: orbit is cleaned, not abandoned

An international cleanup effort removes thousands of large debris objects before cascade thresholds are crossed. The effort costs hundreds of billions, but it preserves trillions in orbital infrastructure. Lieutenant Colonel Sarah Martinez oversees the final mission and declares low Earth orbit safe for future generations. Responsibility replaces denial.

2058: life is found and protected

After two decades of preparation, a sterile probe reaches Europa's ocean. It detects unmistakable signs of life without contaminating the environment. Humanity confirms it is not alone in the solar system and does so without destroying the evidence. Europa is designated a sanctuary, and the discovery reshapes science, philosophy, and culture.

2075: a mature space faring civilization

By mid century, humans live sustainably on the Moon and Mars in limited, carefully chosen settlements. Most of both worlds remain untouched. Asteroid mining occurs under strict regulation, and profits support global development. Technologies developed for space sustainability transform life on Earth, improving water use, food production, and energy systems.

A young Maya Chen, born on Mars, looks at Earth shining in the sky and knows she is there because earlier generations chose patience over speed and stewardship over conquest.

The choice before us

These futures are not predictions. They are warnings and promises grounded in real trends and real decisions. The divergence point is not abstract. It is now. Every launch license, every regulatory vote, every mission timeline pushes us incrementally toward one path or the other.

The question is not whether humanity will go to space. That outcome is already unfolding. The question is whether we will repeat the environmental failures that defined our past or learn from them before exporting those failures beyond Earth.

Two paths lead to the stars. One leads to a sustainable human presence in a protected cosmos. The other leads to collapse, regret, and confinement. The choice is not made by fate or technology. It is made by values, by governance, and by collective will.

Which future do you want, and what are you willing to do to help build it.

17

The Interplanetary Environmental Movement

They said it could not be done.

A grassroots movement to protect space was dismissed as a category error, a confusion of domains that did not belong together. Space policy, people insisted, was the realm of governments, militaries, and billionaires with rockets. Environmental activism worked on Earth because forests could be photographed, oceans could be polluted in visible ways, and air could be breathed or choked. Space was distant, abstract, technical. Let the experts handle it. Let engineers optimize. Let markets decide.

But they were wrong.

In 2019, a teenager sat alone outside a parliament building in Sweden with a hand painted sign about climate change. She was mocked, ignored, and patronized. Within a year, millions of people marched across continents. Greta Thunberg did not invent climate science or environmental ethics.

She did something far more disruptive. She made indifference socially unacceptable. She proved that moral clarity, persistence, and visibility could turn a niche concern into a global movement.

The Interplanetary Environmental Movement began the same way. Not with rockets or treaties, but with questions. Who owns orbit. Who decides what happens to Mars. Who cleans up after satellites fail. Who speaks for worlds with no voices. It began with a handful of scientists frustrated that their warnings were buried in journals, activists who saw the same extractive logic repeating beyond Earth, indigenous leaders pointing out that the sky itself had been treated as empty without consent, and ordinary citizens who refused to accept that inevitability was an excuse for irresponsibility.

They were called unrealistic. Anti progress. Anti space. Too small to matter. They were told that space was too important to regulate and too expensive to protect. They were told to wait. They did not.

They are still small. They are still underfunded. They are still fighting institutions with vastly more money and power. But they exist. They are organizing. They are winning incremental victories. And they might be enough to tip the balance away from the extraction future and toward stewardship.

This chapter is about them. It is about what already exists, what you can do, what institutions must do, and what stands in the way. It is the bridge between imagining a better future and building one.

What is already happening

The most dangerous myth about space environmentalism is that it does not exist yet. In reality, it has been forming quietly for decades, scattered across organizations, disciplines, and countries. What it lacks in scale, it compensates for in seriousness and persistence.

The organizations

The Secure World Foundation was founded in 2002, long before space debris became a mainstream concern. Its focus has always been sustainability, safety, and international cooperation. While it does not organize protests or headline grabbing campaigns, its influence runs through policy briefings, United Nations discussions, and technical working groups. It convenes the annual Space Sustainability Summit, produces widely cited research on debris mitigation, and consistently argues that space is a finite environment rather than an infinite frontier. Its staff is small, its budget modest, but its fingerprints are on many of the conversations that later turn into policy.

For All Moon kind began in 2017 with a narrower goal that proved strategically powerful. Its founders argued that human heritage sites on the Moon, especially the Apollo landing sites, deserved protection. This framing was deliberate. Heritage protection is politically palatable, emotionally resonant, and legally precedented on Earth. By focusing on footprints and artifacts rather than ecosystems, the organization opened the door to the idea that space places could and should be protected. Model legislation inspired by its work has been adopted at state and federal levels in the United States, and its advocacy has reached UNESCO discussions. While heritage protection is not the same as environmental protection, it establishes a critical principle. Some places in space are worth preserving simply because they exist.

The Planetary Society, founded in 1980 by Carl Sagan and colleagues, occupies a unique position. With tens of thousands of members worldwide, it has reach, credibility, and a long history of public education. Its advocacy for planetary defense, science driven exploration, and planetary protection gives environmental concerns a platform that is explicitly pro exploration rather than anti space. When the Planetary Society speaks about Mars contamination or asteroid impacts, it is not dismissed as fringe.

In Australia, the Space Environment Research Centre focuses on the practical side of sustainability. Its work on debris tracking, laser ablation, and mitigation technologies demonstrates that environmental responsibility in space is not just ethical but technical. Problems can be measured. Solutions can be engineered. This matters because policy arguments are far stronger when backed by feasible tools rather than abstract ideals.

Astronomers for Planet Earth emerged in 2020 in response to the rapid deployment of satellite mega constellations. For astronomers, the problem was immediate and personal. Bright satellite trails ruined observations, contaminated data, and threatened the future of ground based astronomy. What began as professional frustration quickly turned into organized advocacy. Thousands of astronomers signed petitions, testified to regulators, and spoke to the media. They reframed the issue from inconvenience to environmental harm, arguing that the night sky itself was being altered without consent.

Alongside these groups, indigenous led initiatives have begun to challenge the assumption that Western technological frameworks have exclusive authority over space ethics. Networks focused on decolonizing space and indigenous astronomy argue that many cultures view the sky as sacred, relational, and alive with meaning. These perspectives bring long term thinking and moral depth into discussions that are often dominated by short term economic logic. Their influence is still limited, but it is growing, particularly among younger scientists and policymakers.

Student organizations also matter. Groups like Students for the Exploration and Development of Space have chapters around the world, and their membership increasingly reflects environmental consciousness. These students will become the engineers, lawyers, and administrators who shape space policy in coming decades. Movements often begin at the margins and grow inward.

Together, these organizations form an ecosystem. They are fragmented, unevenly resourced, and sometimes focused on different aspects of the problem. But they exist, and they are laying the groundwork for something larger.

The victories, small but real

Movements do not begin with sweeping wins. They begin with proof that resistance is possible.

In 2020, astronomers confronted SpaceX over the brightness of Starlink satellites. The initial response from the company was defensive. The satellites were legal. Regulations did not cover brightness. Astronomy would have to adapt. But sustained pressure changed the equation. Scientific papers documented the impact. Professional societies issued statements. Media coverage amplified the concern. Eventually, SpaceX agreed to test darkening methods, including coatings and sun visors. The resulting designs reduced brightness significantly, though not completely. The sky was not saved, but the precedent was established. A private space company altered its design in response to environmental advocacy.

In 2022, the United States Federal Communications Commission shortened the allowable deorbit timeline for satellites from twenty five years to five. This change did not come from nowhere. It followed years of recommendations from NASA, research by debris experts, and thousands of public comments. The rule applies only to new satellites and does not solve the debris problem on its own. But it demonstrates that regulation can move in the direction of responsibility rather than permissiveness.

At the international level, the United Nations Committee on the Peaceful Uses of Outer Space adopted Long Term Sustainability Guidelines in 2019. These guidelines are voluntary, which limits their immediate impact, but they represent consensus among spacefaring nations that sustainability matters.

They provide a shared language and a starting point for future binding agreements. Norms often precede laws.

Within space agencies, planetary protection has held its ground against commercial pressure. The Mars 2020 Perseverance mission adhered to strict contamination standards despite calls to relax them in the name of speed and cost. Science based caution prevailed, reminding policymakers that not all values can be monetized.

There are also early signs of corporate responsibility. Some companies design satellites with proactive deorbit capabilities. Others invest in debris removal as a business model. These actions are not universal, and they are often motivated by self interest as much as ethics. But they show that sustainability and profitability are not mutually exclusive.

Each of these victories is incomplete. None of them solves the larger problem. But movements grow through accumulation, not miracles.

The people behind the work

Movements are sustained by individuals willing to spend years on issues that offer little recognition and slow progress.

Moriba Jah, an engineer and professor, calls himself a space environmentalist. Through projects like AstriaGraph, he has worked to make orbital data open and accessible, arguing that transparency is a prerequisite for accountability. He speaks bluntly about the fragility of orbital space and the absurdity of treating it as infinite. His presence in media and public forums helps translate technical concerns into moral language.

Alice Gorman approaches space from the humanities. As a space archaeologist, she argues that debris and artifacts are not just junk but records of human activity. This reframing challenges the assumption that anything non functional is disposable. It also connects space sustainability to cultural heritage, expanding the coalition of concern.

Aaron Boley's research on the atmospheric effects of satellite reentry adds a critical dimension to the debate. Satellites do not simply disappear when they burn up. They deposit metals and chemicals into the upper atmosphere, with consequences that are only beginning to be studied. This work grounds environmental arguments in hard data rather than speculation.

Aparna Venkatesan bridges astronomy and social justice. She highlights how mega constellations affect both scientific research and indigenous sky traditions, arguing that the commercialization of the sky reproduces colonial patterns of exclusion. Her work insists that ethics cannot be separated from culture.

Timiebi Aganaba focuses on space law and equity. She asks who benefits from space activities and who is excluded. Her scholarship challenges the narrative that space is automatically a global good, pointing out that without deliberate governance, benefits will concentrate among a few.

Jonathan McDowell, a satellite tracker and historian, provides something deceptively simple. Data. By maintaining an independent record of launches, failures, and debris, he gives advocates the numbers they need to counter industry claims. Transparency is power.

Alongside these established figures are students and early career researchers organizing petitions, running social media campaigns, and pushing their institutions to take positions. They bring urgency and impatience that established systems often lack.

What individuals can do

Movements do not grow because everyone becomes an activist. They grow because large numbers of people do small, sustained things.

Educate yourself and others

Awareness is the first bottleneck. Most people still do not know that space debris threatens satellites they use every day or that Mars contamination could erase scientific discoveries forever. Learning is not passive consumption. It is preparation for conversation.

Follow researchers and organizations working on space sustainability. Read reports from groups like the Secure World Foundation. Attend webinars and public lectures, many of which are free. Watch documentaries that explore space ethics alongside exploration.

Then talk. Bring space environmentalism into ordinary conversations. Share articles. Write letters to editors. Explain to children that space is not empty or immune to harm. The modern environmental movement did not grow because everyone read scientific papers. It grew because people talked about polluted rivers and dying forests until those topics became unavoidable.

Support organizations

Most space environmental organizations operate on budgets that would not cover a single rocket launch. Donations, even small ones, matter. Membership numbers matter when organizations lobby governments. Volunteer labor matters when staff are stretched thin.

If you have professional skills, writing, design, translation, data analysis, offer them. If you have time, attend events and amplify their work. If you have money, make recurring donations that allow organizations to plan rather than scramble.

Pressure representatives

Space policy is shaped by elected officials, whether the public engages or not. Contacting representatives works when it is specific and persistent. Ask for support for debris mitigation, planetary protection, and international cooperation. Reference actual bills, hearings, or regulatory processes.

Public comment periods are especially powerful. Regulatory agencies are legally required to consider them, and most receive far fewer submissions than they should. When only industry comments, industry shapes the outcome.

Voting matters. Space policy rarely dominates campaigns, but it is influenced by broader attitudes toward regulation, science, and long term thinking. Ask candidates where they stand. Make it an issue.

Career paths

For students and professionals considering their future, space environmentalism is not a single job but a lens. Space law needs people who understand both orbital mechanics and justice. Planetary protection needs biologists and engineers who value caution. Aerospace engineering needs designers who prioritize sustainability. Policy needs analysts who can translate science into governance.

Science communication is especially critical. Technical knowledge does not change policy unless it is understood. Writers, filmmakers, and educators play a central role in shaping perception.

Consumer and investor power

As consumers, people can reward companies that adopt responsible practices and criticize those that do not. As investors, even small shareholders can push for transparency and accountability. Employee advocacy within aerospace companies can shift internal culture, especially when framed as risk management and long term viability rather than obstruction.

Art and culture

Facts alone rarely move societies. Stories do. Art makes invisible problems visible and abstract stakes emotional. The environmental movement on Earth had its novels, photographs, and films. Space environmentalism needs its own cultural expressions that imagine sustainable futures and warn against reckless ones.

If you create, create about space not just as a backdrop for adventure, but as an environment with limits and meaning. Culture changes what people consider normal.

What institutions can do

Individual action matters, but systemic change requires institutional engagement.

Environmental organizations that focus exclusively on Earth should expand their mandate. The same logic that protects oceans applies to orbit. The same arguments that preserve wilderness apply to Mars. Space is not separate from the environment. It is an extension of it.

Scientific societies should move beyond publishing and into advocacy. When research shows harm, silence becomes complicity. Collective action by scientists carries moral authority that politicians ignore at their peril.

Governments must legislate. Voluntary guidelines are not enough. Environmental impact assessments for launches, binding debris mitigation rules, and strict planetary protection enforcement are all within reach if political will exists. International leadership matters. Treaties do not appear spontaneously. They are built by states willing to push.

Forward thinking companies can lead rather than resist. Adopting high standards early can become a competitive advantage. Sustainability can differentiate brands and reduce long term risk. Industries that race to the bottom often collapse under the weight of their own excess.

The obstacles

Honesty about challenges is essential.

Money is the most obvious barrier. The space industry is measured in hundreds of billions of dollars. Space environmental organizations operate on millions. Lobbying power is asymmetrical. But history shows that money does not guarantee victory when public opinion, science, and moral clarity align.

Complexity is another barrier. Orbital mechanics and contamination protocols are not intuitive. But neither were atmospheric chemistry or ozone depletion until communicators made them understandable. Simplification without distortion is possible.

Urgency is hard to convey when consequences feel distant. The task is to connect space sustainability to everyday life, to show that satellite failure affects navigation, communication, and disaster response.

Coordination is a challenge. Fragmented efforts dilute impact. Building coalitions takes time, trust, and leadership.

Politics complicates everything. Environmental protection has become polarized in ways that obscure shared interests. Framing space sustainability as conservation, stewardship, and responsibility rather than regulation can build broader coalitions. International equity concerns must be addressed honestly through benefit sharing rather than dismissed.

The human thread

Maya Patel did not plan to become an organizer. She studied astrophysics because she loved the sky. During graduate school, a satellite trail ruined months of observational data. At first she was angry at the inconvenience. Then she learned how systemic the problem was.

She joined Astronomers for Planet Earth, helped organize petitions, and sat across from corporate representatives who listened politely and changed little. Small victories kept her going. She left her doctoral program to work full time in advocacy, a decision that strained her finances and relationships.

Years later, exhausted and frustrated, she considered quitting. A mentor reminded her that every movement feels impossible until it is not. She thought of the teenager sitting alone with a sign and decided to keep going.

Her work will not end in her lifetime. She knows that. She also knows that movements are chains, not monuments. Being one link matters.

Ending

The Interplanetary Environmental Movement is not a juggernaut. It is small, underfunded, and outmatched. It will lose battles. It already has.

But it exists. It is growing. It has won enough to prove that space is not immune to public accountability.

This book has shown what is at stake, what futures are possible, and where we stand now. It has shown that doing nothing guarantees the extraction future. Acting gives us a chance at something better.

Not everyone needs to become an organizer. But everyone can contribute. Learn. Share. Support. Speak. Vote. Create. Care.

Movements are not built by heroes. They are built by millions of people deciding that an issue matters enough to spend a little time, a little money, and a little attention.

The question is no longer whether a movement to protect space can exist. It already does. The question is whether it will grow large enough, fast enough, to matter.

That depends on you.

History is watching. The cosmos is waiting. The choice is still open.

18

Epilogue

2045: A Reckoning

Mars Historical Archive, 2045.

The circular chamber was quiet except for the low hum of the life support system and the faint vibration of artificial gravity. Beyond the reinforced windows, the Martian dawn crept slowly across the horizon, a pale apricot light filtering through dust suspended in the thin air. The planet looked ancient and patient, as if it had seen far worse than humanity's brief presence and would see far more after it.

Dr. Elena Vasquez stood at the center of the room, hands folded behind her back, facing a semi circle of young people seated on tiered benches. They were between sixteen and twenty two years old, born between 2023 and 2029, members of the first generation that could not remember a world without constant discussion of space debris, Mars missions, and orbital congestion. To them, space had always been crowded, contested, and controversial.

Above Elena, two vast holographic displays hovered in the air. One glowed a muted red, labeled Timeline A. The other shimmered blue and green, labeled

Timeline B. Destruction and Stewardship. The same images she had shown countless groups before, rendered with exquisite detail. Orbital debris clouds thickening like smog. Pristine craters preserved beneath careful domes. Scarred moons. Protected sanctuaries.

The question hung in the air, heavy and unspoken.

Which one did we choose.

The young people leaned forward, some clasping their hands, others resting their elbows on their knees. They had lived through the early chapters of the story without realizing it was a story at all. They had seen headlines scroll past their screens, had watched launches live streamed, had absorbed arguments about regulation and innovation without ever seeing the whole arc. They knew pieces. They did not know the pattern.

Elena did.

She was seventy eight now, born on Earth in 1967, raised under a sky still mostly untouched by satellites, educated during the first great environmental reckoning of the twenty first century. She had moved to Mars in 2038, not because she believed in escape, but because she believed history should be witnessed where it was being made. She had lived through the critical decade, the years between 2025 and 2035 when choices hardened into trajectories. She had watched decisions made in committee rooms ripple outward into space itself.

She knew the answer.

It was not the answer they expected.

The honest assessment

Elena raised her hand, and the holograms shifted. The sharp contrast between red and blue softened, the images blending and rearranging into a more complicated mosaic.

"You want to know which timeline we are in," she said, her voice steady, amplified gently by the room. "Timeline A or Timeline B. Destruction or Stewardship."

She paused, letting the anticipation build, then continued.

"The truth is neither. And both.

"We are in Timeline C. The messy, compromised, partially successful, partially failed reality that happens when millions of people make millions of decisions, some wise, some foolish, most somewhere in between. Let me show you what actually happened."

The victories, real but incomplete

The first images that appeared were orbital maps of Earth, layers of colored dots encircling the planet like nervous systems.

"Between 2025 and 2038," Elena said, "we came closer to losing low Earth orbit than most people ever realized. The risk of cascading collisions was not theoretical. It was imminent."

In 2027, after years of warnings and a series of near misses that finally broke through political inertia, the International Space Cleanup Initiative was launched. It was not as ambitious as the most optimistic proposals had demanded, but it was real. One hundred eighty billion dollars were committed, fifteen nations agreed to participate, and for the first time, active debris removal moved from pilot projects to sustained operations.

Over the next eight years, more than eight thousand large debris objects were removed from orbit. Harpoons, nets, laser ablation systems, and drag augmentation devices worked in concert, an awkward but effective choreography. At the same time, new satellites were required to carry improved collision avoidance systems, and constellation designs became marginally less reckless.

In 2036, a collision between two large objects nearly triggered the cascade everyone feared. For forty eight hours, tracking centers around the world coordinated emergency maneuvers, sharing data that would once have been guarded for national security reasons. The worst case scenario was avoided by margins measured in meters and seconds.

"We did not get a clean orbit," Elena told the students. "We got a survivable one. We did just enough, just in time. Barely."

The hologram shifted to Mars, its surface overlaid with colored zones.

Mars protection was similarly imperfect. In 2030, an International Mars Compact was signed, not a full treaty with strong enforcement, but an agreement strong enough to matter. Sixty percent of the planet was designated protected, including Hellas Basin and other regions with high scientific value. It was less than scientists had hoped for and more than industry had wanted.

The first crewed landing in 2033 touched down in a designated green zone, with contamination protocols that were followed carefully but not flawlessly. Human error and equipment failures led to localized contamination, the inevitable result of placing fragile biological systems into an alien environment. In 2037, a controversial drilling operation in a yellow zone sparked international outrage, but even then, red zones remained untouched.

In 2042, subsurface fossil microbes were discovered, evidence of ancient Martian life that had gone extinct nearly three billion years earlier. The discovery was possible precisely because the most sensitive regions had been protected.

"We did contaminate parts of Mars," Elena said. "That is a failure we will live with forever. But we also preserved the places that mattered most. Because of that, we learned that Mars once lived."

The story of mega constellations followed the same pattern. Explosive growth in the mid 2020s damaged astronomy irreversibly, but not fatally. An international coordination agreement in 2028 capped satellite numbers, imposed brightness restrictions, and established an orbital slot allocation system. The night sky was never the same, but it was not erased entirely. Ground based astronomy adapted, and space based telescopes expanded.

Europa, by contrast, stood as a rare example of patience rewarded. After the Europa Clipper mission arrived in 2027, a fierce debate erupted over whether to land immediately or wait. The decision to delay was costly and unpopular in some quarters, but contamination free drilling technology matured by 2035, and a carefully planned lander was approved for a 2048 mission. Europa remained untouched, its ocean still a question rather than a casualty.

"We learned," Elena said softly. "Not quickly. Not easily. But we learned."

The failures, real and painful

The hologram darkened, showing the Moon's south pole.

"We also failed," Elena continued. "In ways that cannot be undone."

Lunar mining began in 2031, driven by water ice demand and geopolitical competition. Shackleton Crater's rim was eventually protected after intense lobbying, but surrounding craters were not. By the time public outrage reached critical mass, scars were already visible through amateur telescopes. Scientific records billions of years old were lost before anyone fully understood their value.

Mining practices improved after the damage was done, and sixty percent of the south pole region remained pristine, but the loss was permanent.

"We scarred the Moon," Elena said. "And no future wisdom can unscar it."

The asteroid belt followed a familiar frontier pattern. Early regulation was weak, enforcement weaker. Profits flowed faster than caution. Environmental assessments existed, but compliance varied widely. The situation remained manageable, but precarious, with long term sustainability still an open question.

Orbital debris, despite cleanup efforts, never truly went away. New debris replaced old. Management became a permanent expense rather than a one time fix. In 2041, a minor cascade in geostationary orbit destroyed twenty satellites before it was contained. The lesson was clear. Vigilance could not lapse.

"We are holding the line," Elena said. "But holding it costs money, energy, and attention. One mistake could undo decades of work."

The current state, 2045

Elena pulled up live data, numbers and maps updating in real time.

Low Earth orbit was crowded but stable, with forty two thousand active satellites and an ongoing cleanup budget of eight billion dollars a year.

The Moon hosted four scientific stations and a partially protected south pole. Mars supported three settlements, small and carefully managed, with contamination localized rather than planetary. The asteroid belt remained uncertain. Europa waited.

Earth itself showed signs of recovery, aided by space based solar power and closed loop technologies developed for off world survival.

"We are not in catastrophe," Elena told the students. "And we are not in paradise. We are surviving. And survival, in this context, is an achievement."

Section 2: Why we did not get Timeline B

The students listened as Elena explained the gap between vision and reality.

In the critical decade, treaty efforts succeeded only partially. A Space Environment Treaty was proposed in 2026, negotiated fiercely, and ratified in a weakened form by eighty nine nations in 2029. It created binding obligations, but left loopholes large enough for determined actors to exploit.

Public pressure grew from thousands to millions, but not to the overwhelming scale required to dominate political agendas crowded by climate crises, pandemics, and economic instability. The movement mattered, but it was not irresistible.

Technological progress delivered tools, but funding constrained deployment. Cultural shifts began, but did not fully mature. Political will appeared in bursts rather than as a sustained force.

"We achieved about sixty percent of what Timeline B required," Elena said. "That is why we are sixty percent of the way there. And sixty percent was enough to avoid collapse."

She looked at the students, her expression firm.

“But enough to survive is not the same as enough to thrive.”

What the next generation must do

“You will inherit this,” Elena said. “All of it. The frameworks, the technologies, the scars, and the unfinished work.”

By 2050, debris management would test its limits again. Mars would face pressure to expand settlements. Asteroid mining would demand stronger governance. The outer solar system would tempt humanity with worlds even more fragile than Mars.

“We are handing you tools and problems,” she said. “We did not solve everything. We kept the door open.”

Her voice softened.

“Your task is to push Timeline C closer to B. To strengthen what works. To close loopholes. To expand protection before expansion overwhelms it.”

The final message

Elena smiled, almost apologetically.

“I am not really here,” she said. “I am a device. A way to help you, reading this in 2025, imagine the consequences of your choices.”

The message, she emphasized, was real. The choice was real.

Reality would never match a clean timeline. It would always be contested. But direction mattered.

"You now know what is at stake," she said. "What you do with that knowledge is up to you."

Care, she reminded them, was the first step. Action followed from care.

The last words

One student finally raised a hand.

"So did we win or lose?"

Elena smiled, a mixture of sadness and pride crossing her face.

"Neither," she said. "We are still fighting. And that means we have not lost."

She turned off the holograms. The two timelines faded into nothing.

Outside, Mars stretched toward the horizon, scarred in places, untouched in others. Above, Earth rose in the thin sky, blue and fragile, still alive.

Both worlds waited.

The future was not written.

It was being written.

www.ingramcontent.com/pod-product-compliance
Lightning Source LLC
LaVergne TN
LVHW090400210926
848480LV00002B/409

9798224306503